Sodium Explosion Critically Burns Firefighters Newton, Massachusetts

Investigated by: J. Gordon Routley

This is Report 075 of the Major Fires Investigation Project conducted by TriData Corporation under contract EMW-4-4329 to the United States Fire Administration, Federal Emergency Management Agency.

FEMA

Department of Homeland Security
United States Fire Administration
National Fire Data Center

U.S. Fire Administration Fire Investigations Program

The U.S. Fire Administration develops reports on selected major fires throughout the country. The fires usually involve multiple deaths or a large loss of property. But the primary criterion for deciding to do a report is whether it will result in significant "lessons learned." In some cases these lessons bring to light new knowledge about fire--the effect of building construction or contents, human behavior in fire, etc. In other cases, the lessons are not new but are serious enough to highlight once again, with yet another fire tragedy report. In some cases, special reports are developed to discuss events, drills, or new technologies which are of interest to the fire service.

The reports are sent to fire magazines and are distributed at National and Regional fire meetings. The International Association of Fire Chiefs assists the USFA in disseminating the findings throughout the fire service. On a continuing basis the reports are available on request from the USFA; announcements of their availability are published widely in fire journals and newsletters.

This body of work provides detailed information on the nature of the fire problem for policymakers who must decide on allocations of resources between fire and other pressing problems, and within the fire service to improve codes and code enforcement, training, public fire education, building technology, and other related areas.

The Fire Administration, which has no regulatory authority, sends an experienced fire investigator into a community after a major incident only after having conferred with the local fire authorities to insure that the assistance and presence of the USFA would be supportive and would in no way interfere with any review of the incident they are themselves conducting. The intent is not to arrive during the event or even immediately after, but rather after the dust settles, so that a complete and objective review of all the important aspects of the incident can be made. Local authorities review the USFA's report while it is in draft. The USFA investigator or team is available to local authorities should they wish to request technical assistance for their own investigation.

This report and its recommendations were developed by USFA staff and by TriData Corporation, Arlington, Virginia, its staff and consultants, who are under contract to assist the Fire Administration in carrying out the Fire Reports Program.

The USFA greatly appreciates the cooperation received from Fire Chief Joseph S. Daniele and Captain Joseph La Croix of the Newton, Massachusetts, Fire Department; Trooper James Bradbury of the Office of the State Fire Marshal; and Steve Coan, Director of the Massachusetts Firefighting Academy.

For additional copies of this report write to the U.S. Fire Administration, 16825 South Seton Avenue, Emmitsburg, Maryland 21727. The report is available on the USFA Web site at http://www.usfa.dhs.gov/

U.S. Fire Administration
Mission Statement

As an entity of the Federal Emergency Management Agency (FEMA), the mission of the U.S. Fire Administration (USFA) is to reduce life and economic losses due to fire and related emergencies, through leadership, advocacy, coordination, and support. We serve the Nation independently, in coordination with other Federal agencies, and in partnership with fire protection and emergency service communities. With a commitment to excellence, we provide public education, training, technology, and data initiatives.

TABLE OF CONTENTS

Sodium Explosion Critically Burns Firefighters
Newton, Massachusetts
October 25, 1993

Local Contact: Fire Chief Joseph S. Daniele
                   Newton Fire Department
                   1164 Newton Street
                   Newton, MA 02159
                   (617) 552-7272

OVERVIEW

Eleven firefighters were burned, six seriously, one critically, and one extremely critically, in an explosion that occurred while they were attempting to extinguish a sodium fire in a metals processing establishment in Newton, Massachusetts, on October 25, 1993. The incident illustrates how quickly and unpredictable an apparently minor hazardous materials situation can change, with tragic consequences. It also shows how standard protective clothing and equipment, designed for structural firefighting, is dangerously inadequate for a molten metals incident.

The situation was caused by a deviation from standard procedures for handling waste sodium at the facility. The incident provides an important series of lessons for all firefighters on the risks involved with sodium and other flammable metals and on the need to obtain reliable information from responsible individuals at hazardous materials incidents.

Fires in sodium and other waste reactive metals are uncommon, and the circumstances of this incident are particularly unusual. Previous experience and pre-fire planning at the facility contributed to a false sense of security among the firefighters, who believed that the incident could be handled easily and without significant risk. Employees at the facility did not provide information that would have caused the officers in charge of the incident to more fully evaluate the risks of this particular situation before initiating action.

SUMMARY OF KEY ISSUES

Issues	Comments
Cause of Explosion	Employees deviated from standard procedures by attempting to dispose of an excessive amount of residual sodium. The sodium overflowed and came in contact with water causing an explosion and fire.
Second Explosion	A second explosion occurred when firefighters were attempting to extinguish the residual fire from the first explosion. The firefighters were splashed with burning molten sodium.
Casualties	Eleven firefighters burned, two very critically, six others seriously. Two plant employees were also injured.
Risk Assessment	The hazards of burning liquid sodium exceed the capabilities of the fire department. There is an extreme risk of explosion, extinguishing agents are ineffective, and protective clothing is inadequate.
Additional Hazard	The use of the same enclosure to perform wet washing and to burn-off excess sodium created an unnecessary hazard. Sodium should never be handled in a location where there is any possibility of contact with water.
Action Plan	Responding fire department personnel were not provided with essential information that should have been considered in the development of an action plan. A full evaluation of the risks and potential consequences of this incident would have resulted in the conclusion that the safe plan would have been to take no action.
Structural Protective Clothing	Structural protective clothing are self-contained breathing apparatus and are not designed to provide adequate protection for exposure to molten metal. There is no practical protective clothing for this hazard.
Proper Use of Protective Clothing	Members involved in the incident would have been better protected if they had been wearing full protective clothing ensembles that meet current standards and had used the chin straps on their helmets, pulled up 3/4 length boots, and (in one case) not worn a turnout coat without the liner.
Communications	With ambulance radios set to police (dispatch) channel, heavy radio traffic interfered with the Incident Command getting them into the scene to transport burn victims.
Molten Metal	Burns caused by molten metal are more severe than other types of burns, because the metal is extremely hot and impregnates protective clothing.

The objective of the United States Fire Administration, in preparing this report, and of Chief Joseph Daniele of the Newton Fire Department in requesting the participation of the U.S. Fire Administration in the investigation, is to share the lessons that were learned so that similar painful and tragic situations can be avoided in the future. The actions that were taken prior to and during this incident have been analyzed in great detail to determine what went wrong.

The analysis, which was conducted with the luxury of time and access to all available information and expertise, indicates that the action that was taken involved a high level of risk and resulted in the situation that is described. It must be recognized that these resources were not available to the individuals who had to face the situation as it was presented to them at the time. This report should not be interpreted as a criticism of the decisions that were made or the actions that were taken.

NEWTON

Newton is a suburb located immediately to the west of Boston, Massachusetts. Newton is primarily a residential community with approximately 80,000 residents, although it has significant commercial and business areas and a relatively small industrial area. The community includes some very affluent areas, as well as a mixture of other residential neighborhoods. The main campus of Boston College and several smaller educational institutions are also located in Newton.

The Newton Fire Department has 199 career personnel, operating seven engine companies and three truck companies, with an assistant chief on duty at all times. All of the companies operate with four member crews during the winter months; however, the engine companies often operate with three crew members in the warmer months.

The Newton Fire Department is headed by Chief Joseph S. Daniele and includes its own Fire Prevention Bureau and Communications Center. Newton takes an active part in Metro Fire, the mutual aid system that covers Boston and 35 surrounding jurisdictions. The Newton Fire Communications Center serves as the primary mutual aid coordinating center for the Metro Fire system. Newton also participates in the regional hazardous materials team with several of the other Metro Fire jurisdictions. The Boston Fire Department operates an additional Hazmat team to serve the central city, and the two teams provide backup for each other.

Emergency medical service in Newton is provided by a private ambulance company, under the direction of the police department. Advanced life support ambulances are stationed at two fire stations but are dispatched by the police department. The fire department does not routinely respond to EMS calls, unless they involve rescue or extrication. The ambulance company serves several communities in the area and has units deployed at several additional locations around Newton.

LOCATION OF THE INCIDENT

The facility where the incident occurred is located in an industrial area in the southern part of the city. The complex consists of several one story brick industrial buildings on a crowded site. The particular building where the incident occurred is toward the back of the complex. The site plan and building plan appear on the following pages. Several "high tech" companies have their manufacturing and processing facilities in the immediate area.

The incident occurred at a plant operated by H. C. Starck, Inc., a multinational company that produces a wide range of products. The facility in Newton produces items manufactured from tantalum, a rare metal that is used for components that require exceptionally high strength and temperature resistance. The tantalum parts are used in assemblies such as jet engines and nuclear reactors.

Tantalum is shipped to the H. C. Starck facility as a salt, tantalum chloride, which must be converted to an extremely high grade metallic state before it can be molded and machined into finished products. Sodium is used in the process as a reducing agent; the tantalum is converted from the salt compound to a pure metal and the metallic sodium is converted to a non-hazardous salt compound (sodium chloride). The reaction takes place in a closed system and all of the by-products are retained for other industrial uses.

H.C. STARCK INC. – SITE PLAN
NEWTON, MASSACHUSETTS

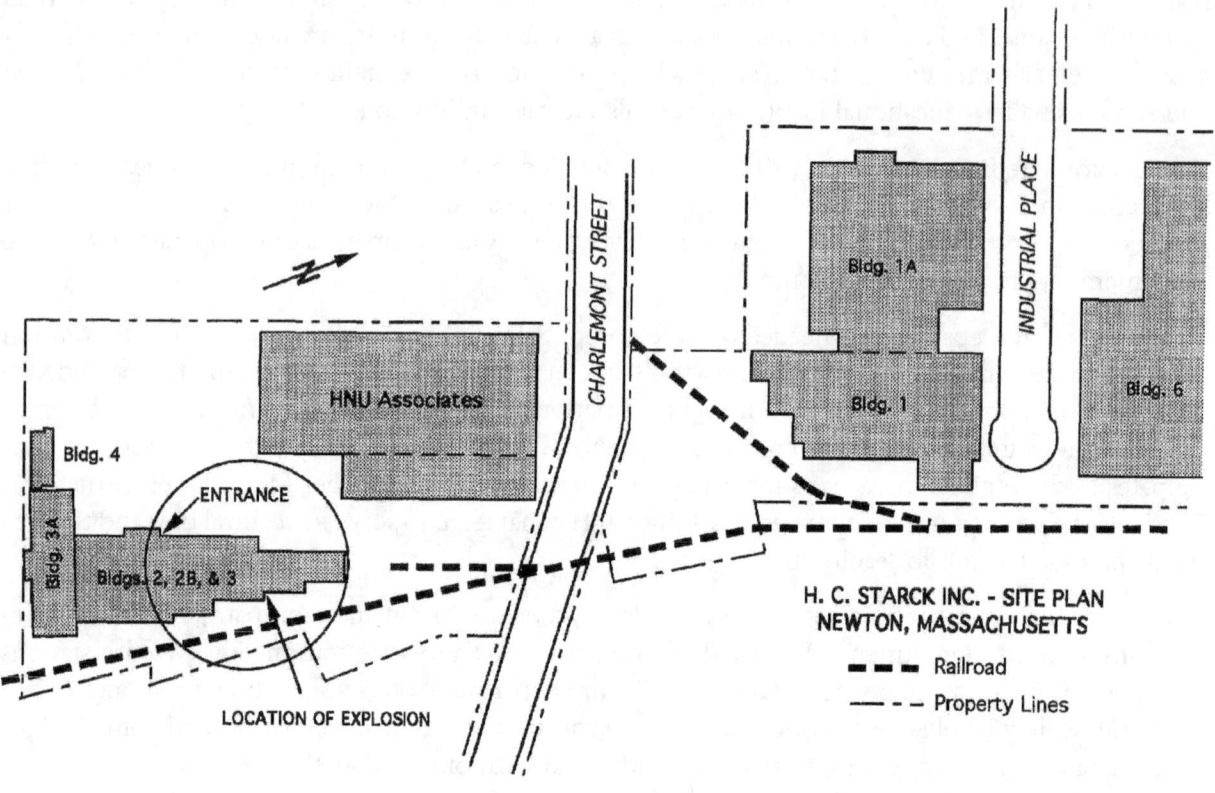

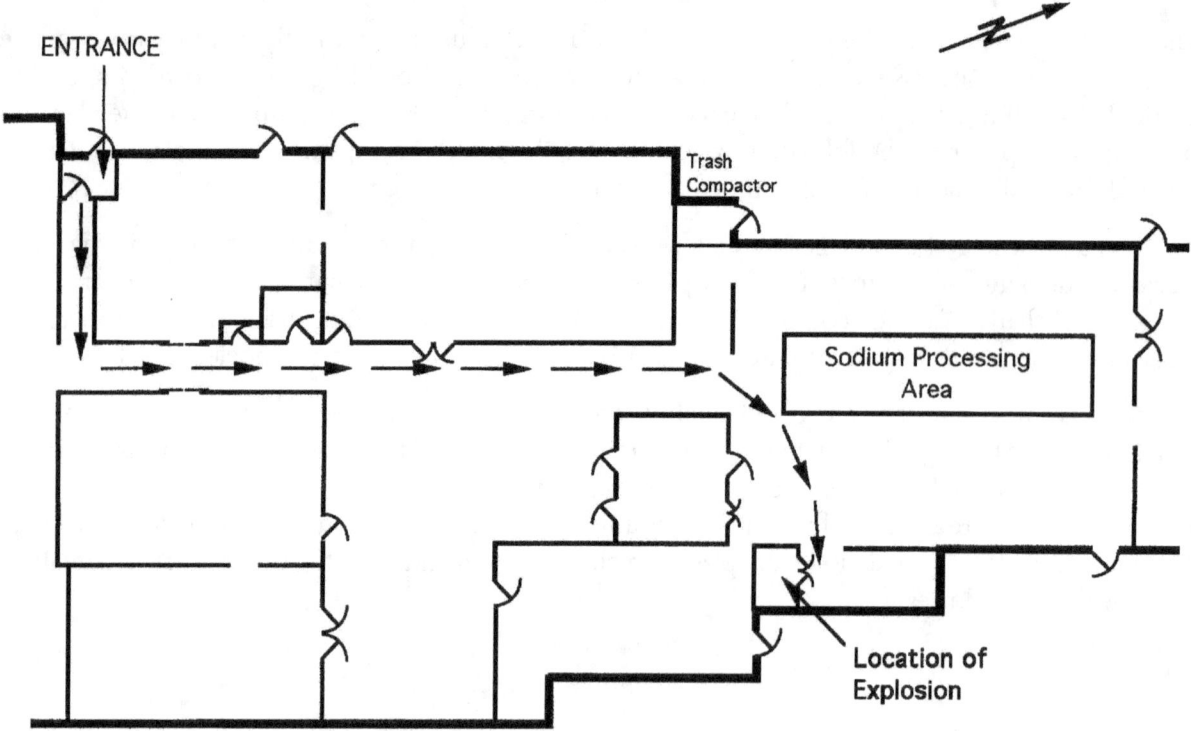

The facility is reported to have an exceptionally good relationship with the Newton Fire Department. The plant has a full-time safety specialist, who maintains a close liaison with the Newton Fire Department, and the company is involved with the Local Emergency Planning Committee for hazardous materials incidents. All of the Newton Fire Department companies that respond to the plan had been given familiarization tours and were provided with information on the hazards of materials used at the plant, including extinguishing procedures for minor sodium fires. Employees are also trained to handle minor sodium fires.

There had been some incidents involving sodium at the facility in the past. The Newton Fire Department had responded to some of these incidents, which were handled with no unusual problems. Large containers of sodium chloride are located in areas where sodium is handled for use as an extinguishing agent.

SODIUM

Sodium is classified as a hazardous material, primarily because of its extreme reactivity when it comes in contact with water and many other substances.[1] (See Appendix A for the Sodium MSDS Sheet and other material on its properties, uses, storage, and handling.) Because of its reactivity, sodium is seldom encountered in the pure metallic state, except when it is being used in an industrial process or for some extremely special application. It must be shipped in sealed containers, because it will react even with the moisture in the air on a humid day.

Sodium is a constituent of several non-hazardous compounds that are in common use, such as table salt (sodium chloride) and several pharmaceutical products. The metallic form of sodium is used in applications that require unusual heat transfer and electrical conductivity properties. Pure sodium is encountered more frequently, however, as an agent in the processing of other substances. It is an extremely powerful reducing agent, with the ability to strip oxygen atoms and other atoms or molecules from otherwise stable molecules. These reactions usually release large amounts of energy and additional chemical by-products are often created, some of which are hazardous in themselves.

Sodium is most widely recognized for its violent reaction with water. Pure sodium will break apart water molecules, separating the hydrogen atoms from the oxygen atoms. The sodium combines with the oxygen and liberates the hydrogen. The oxidation of the sodium is a combustion process, in which the sodium burns with a yellow flame to produce an ash (sodium oxide), which is liberated as a dense white acrid smoke. The hydrogen is released as a gas, which usually explodes in the air as the hydrogen recombines with oxygen from the ambient atmosphere.

In addition to creating sodium oxide and hydrogen gas, contact with moisture can create sodium hydroxide, a corrosive liquid, which can cause corrosion burns to exposed skin.

The power of sodium to break apart other compounds that contain oxygen atoms and/or atoms with similar properties to oxygen make it an extremely valuable reducing agent with numerous applications in the processing of other materials. Sodium is a solid at normal ambient temperatures

---

[1] El DuPont is the only remaining producer of sodium in the United States. The demand for sodium has decreased significantly in recent years due to the switch from regular gasoline to unleaded gasoline as a motor fuel. Sodium was used in the production of tetraethyl lead, an important additive in regular gasoline. Sodium is also used as a coolant in some nuclear reactors, and particularly in experimental breeder reactors. For additional information on sodium, readers should contact El DuPont at (800) 441-9372.

but melts at the relatively low temperature of 208 degrees Fahrenheit. Above 208 degrees, it can be transferred and mixed with other substances as a liquid; however, it must be kept in a closed system because it will auto-ignite in air at temperatures only slightly above its melting temperature. Liquefied sodium flows easily, with a viscosity similar to water.

Once ignited, sodium is very difficult to extinguish. It will react violently with water, as noted previously, and with any extinguishing agent that contains water. It will also react with many other common extinguishing agents, including carbon dioxide and the halogen compounds and most dry chemical agents. The only safe and effective extinguishing agents are completely dry inert materials, such as Class D extinguishing agents, soda ash, graphite, diatomaceous earth, or sodium chloride, all of which can be used to bury a small quantity of burning sodium and exclude oxygen from reaching the metal.

The extinguishing agent must be absolutely dry, as even a trace of water in the material can react with the burning sodium to cause an explosion. Sodium chloride is recognized as an extinguishing medium because of its chemical stability; however, it is hydroscopic (has the property of attracting and holding water molecules on the surface of the salt crystals) and must be kept absolutely dry to be used safely as an extinguishing agent. Every crystal of sodium chloride also contains a trace quantity of moisture within the structure of the crystal.

Molten sodium is <u>extremely</u> dangerous because it is much more reactive than a solid mass. In the liquid form, every sodium atom is free and mobile to instantaneously combine with any available oxygen atom or other oxidizer, and any gaseous by-product will be created as a rapidly expanding gas bubble within the molten mass. Even a minute amount of water can create this type of reaction. Any amount of water introduced into a pool of molten sodium is likely to cause a violent explosion inside the liquid mass, releasing the hydrogen as a rapidly expanding gas and causing the molten sodium to erupt from the container.

When molten sodium is involved in a fire, the combustion occurs at the surface of the liquid. An inert gas, such as nitrogen or argon, can be used to form an inert layer over the pool of burning liquid sodium, but the gas must be applied very gently and contained over the surface. Except for soda ash, most of the powdered agents that are used to extinguish small fires in solid pieces or shallow pools will sink to the bottom of a molten mass of burning sodium – the sodium will float to the top and continue to burn. If the burning sodium is in a container, it may be feasible to extinguish the fire by placing a lid on the container to exclude oxygen.

Most municipal fire departments rarely, if ever, come in contact with pure sodium, particularly molten sodium, in any significant quantities. It is shipped in sealed containers and can only be used under extremely controlled conditions in closed industrial processes. It is most often used within large industrial complexes, where municipal fire departments are unlikely to become involved with it. Industries that use sodium must be extremely careful with it, because of the consequences of using it unsafely; many have plant fire brigades trained to handle small sodium incidents.

It is also used in high energy/high temperature systems as a heat transfer medium. In this application it may be encountered at some nuclear power facilities and in experimental installations that are involved in high energy power generation and transmission.

DISPOSAL OF WASTE SODIUM

The pure sodium is shipped to the plant in Newton in 55 gallon steel drums each of which contains 400 lbs. of reactor grade (more than 99.9 percent pure) sodium metal in a fused (solid) state. The solid mass of sodium occupies all but a few inches at the top of the drum and the remaining space is filled with argon, an inert gas, to avoid contamination of the product in transit or storage.

To extract the sodium, the steel drum is placed in a special heater jacket that warms the contents until the sodium liquefies. A bayonet pick-up tube is then inserted into the drum and additional inert gas is introduced at the top of the drum, to displace the liquefied sodium. The sodium is drawn up into the tube and directly into the closed processing system. The entire process takes place in a sealed atmosphere, since even the moisture and other contaminants in the outside ambient air can contaminate the sodium, creating a risk of explosion and making it unusable for the process.

The normal procedure is to extract all of the sodium from the drum, sometimes over a period of several days, down to the point that it can no longer be drawn into the pick-up tube. This usually leaves two to five pounds of sodium at the bottom of the drum which is allowed to cool and solidify after the tube is withdrawn. The residual sodium, referred to as a "heel," usually occupies less than 1/2 inch at the bottom of the drum. This amount of sodium is impractical to extract or recycle; burning-off is a common practice in the industry to dispose of small quantities of residual sodium.

Burning-off the residual sodium avoids having to dispose of the drums as a hazardous waste. This is practical, because the sodium can be completely consumed and converted to sodium oxide ash, which is collected by a filter system and can be disposed of much more easily. The plant routinely has several drums with residual sodium to burn off each week. The task is usually performed during the second or third shift.

The waste sodium is burned-off in a special enclosure that was built for cleaning drums. The room is approximately 8 ft. x 10 ft., with an exhaust system built into the roof to draw out the smoke. The burning sodium produces large quantities of smoke which is passed through a scrubber system to capture the sodium oxide ash and prevent atmospheric contamination. The room has concrete block walls, with a large blowout explosion relief panel in the exterior wall to allow the force of an explosion to be vented to the outside. The blowout panel consists of a plywood sheet, held in-place by a metal framework.

The access to the enclosure is a double doorway opening into a maintenance area of the plant. The doors are "blast doors" reinforced to stay closed while the force of an explosion is directed to the outside through the blowout panel, instead of through the doorway to the interior of the plant.

In addition to being designed for the disposal of waste sodium, the room was also designed for wet washing drums and equipment that had contained other products and other items used in the process. Spray nozzles are located in one part of the room and the floor is a heavy metal grate, which allows any runoff to drain into a system of shallow troughs that lead to a holding tank and waste treatment system. Drums and equipment can be placed in the room and flushed, with all runoff draining down through the grates and into the troughs. There is often residual water under the floor grates, after the room has been used for wet washing.

To burn-off a sodium drum, the near empty drum is placed in a special cradle that sets it at a slight angle inside the room (diagram appears on the following page). The cradle has a rocker end to allow the drum to be tilted from a vertical position to an almost horizontal position. A metal drip pan is

SODIUM DISPOSAL - NORMAL OPERATIONS
2 - 5 lb. of Sodium

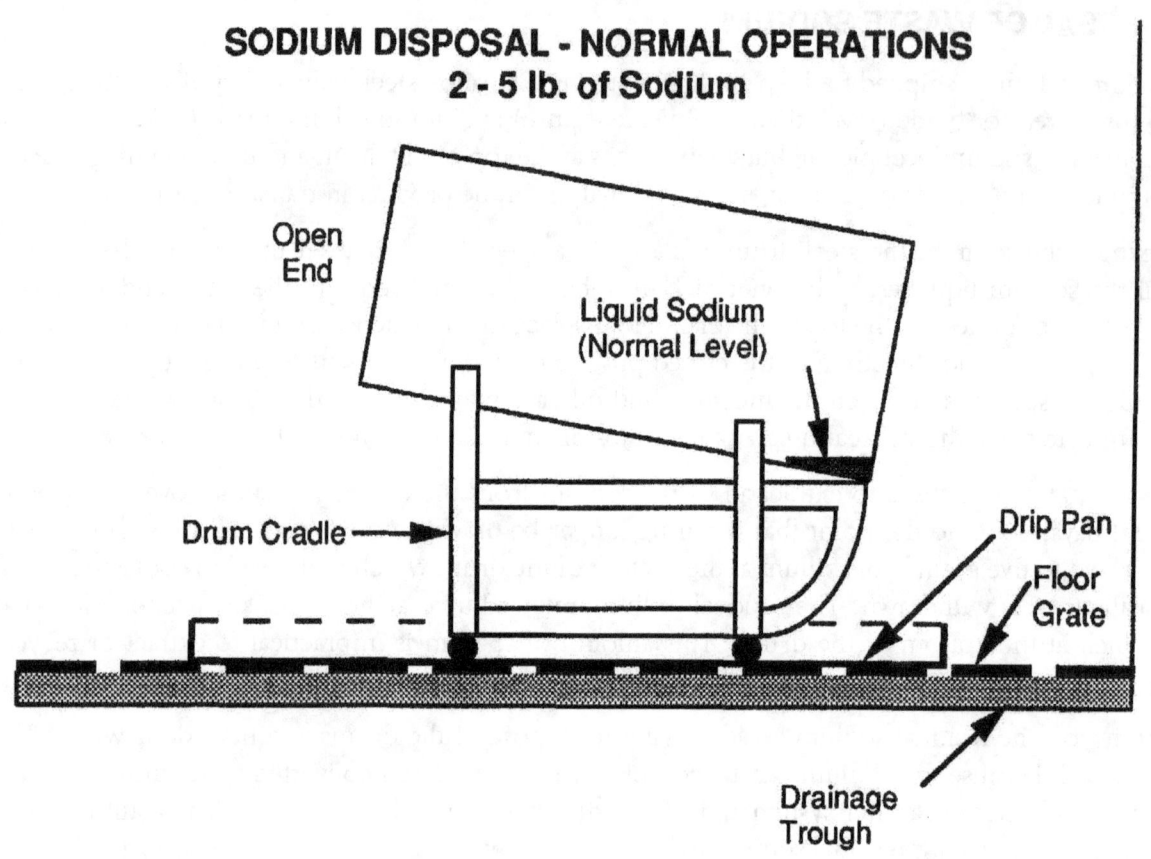

SODIUM DISPOSAL INDICATING OVERFLOW
(with 100 lb. of Sodium)

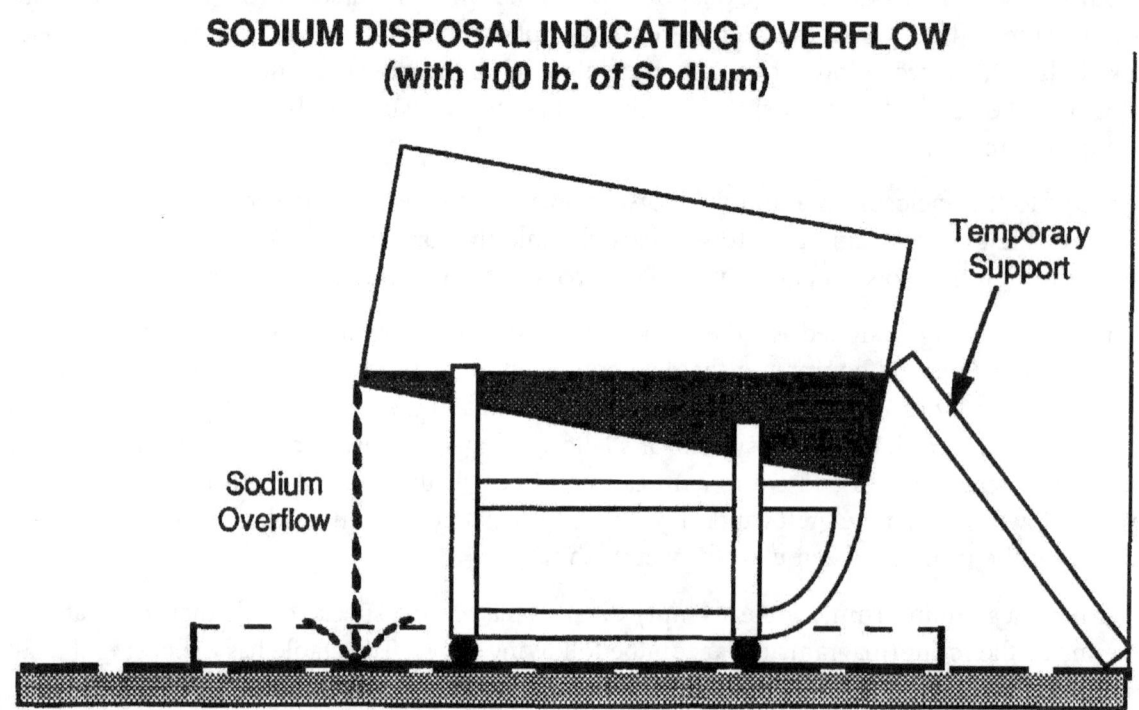

placed under the drum cradle to catch anything that might drip out of the drum and keep it from contacting the floor grates or falling into the troughs. The top of the drum is completely removed and the drum is positioned with the open end facing toward the blast doors.

After the blast doors are closed and secured with a vertical bar, a worker ignites the sodium with a MAPP gas lance, which is inserted through an access hole in the blast doors and directly into the drum. Any residual sodium coating the sides of the drum is melted and forms a pool at the low end of the drum. When the sodium is ignited, the lance is removed and the product is allowed to burn itself out, which usually takes one to two hours. The sodium first melts and form a puddle in the low end of the drum, then the liquid usually burns calmly with a glowing red combustion.

The burning sodium sometimes sputters and emits small flares of glowing metal which are contained within the room. A heavy gray/white smoke is produced, which is drawn immediately up into the exhaust system and out through the scrubbers. When the sodium is completely burned out and the drum has cooled, it can be removed and sent out for disposal as a non-hazardous waste.

INCIDENT HISTORY

On the afternoon of October 25, 1993, a drum in the heater jacket was allowed to cool before all of the sodium had been extracted. Due to a mathematical error by the operator, approximately 100 lbs. of sodium, almost one quarter of the drum's capacity, was left in the drum when the heaters were turned off. The re-solidified mass of sodium could not be reheated and drawn into the processing system, because it had been contaminated by contact with outside air.

The company's standard operating procedures contained detailed specific instructions on how to burn-off the sodium residue from a drum, anticipating that less than 10 pounds of residue would have to be burned off. There was no reference to guide employees on disposal of a large heel of sodium.

Similar situations were reported to have happened occasionally in the past and caused major problems for the employees responsible for disposal of the drums. On one occasion, several months earlier, they had used a chipping hammer to chip the sodium out of the barrel and burn it off in small quantities. The chipping-out process was labor intensive and time consuming and, when the sodium was chipped out, the chips came in contact with the sweaty skin of the worker. This caused painful burns as the sodium reacted with his perspiration. They had decided to do something different if it happened again. The alternatives are to dispose of the drum as hazardous waste or to return it to the original supplier for recycling.

On the night of October 25, the night shift supervisor advised the worker who handled the drums that they had another drum with a large quantity of residual sodium and they were going to try burning it off, following the normal disposal procedure. They anticipated that it would take longer, possibly all night to burn this quantity of sodium.

The room had been used earlier in the evening to wash a reactor head and the floor grates were reported to be damp when the drip pan was placed on top of them. The pan, which is leaned against the wall of the room when it is not in use, is also believed to have been wet when it was laid on top of the floor grates and there may have been water under the grates.

The drum was removed from the process area and transported to the disposal room where the top was removed. When it was attempted to place the drum in the cradle, the mass of solid sodium at the bottom would not allow it to sit in the normal near-horizontal position; a metal agitator shaft was used as a strut to keep it from tilting back to the upright position. The braced drum was placed in the normal position for burning, on top of the drip pan with the open end toward the door.

Shortly after the sodium was ignited with the MAPP gas lance, the worker looked through a viewing port and noted that the fire was creating much more smoke than usual. He also noted within a few minutes that the solid sodium was liquefying and creating a liquid pool that was quickly filling the deep end of the drum. He became concerned and summoned the supervisor who had directed him to burn the drum in this manner.

Moments later, as the supervisor and two employees were discussing possible courses of action, they noted that the liquid level was almost up to the lip of the drum at the open end and the heavy smoke was filling the burnroom. A few second later an explosion occurred inside the burnroom that shook the area and knocked them off their feet. Although the blast doors held closed, some flecks of burning sodium were expelled through the openings in the door and struck two of the plant personnel, causing minor burns. White smoke filled the immediate area and began to spread to other parts of the building. All three employees evacuated the area, assisted by other plant personnel. The injured personnel were treated for minor burns by other employees.

FIRE DEPARTMENT RESPONSE

The Newton Fire Department received box 6237, the alarm box located at the front door of the building, at 2131 hours. As the box was being transmitted, the first of several telephone calls was received, reporting an explosion and fire with injuries at the plant. The details given by callers were incomplete and conflicting; however, the assignment was upgraded from the regular response (two engines, one truck, and a command officer) to the full building fire response, which adds an engine company from the neighboring town of Needham. An additional Newton engine company was also dispatched as the 4th due engine company on the full box assignment. All responding units were advised of the reports of a fire and explosion with injuries.

The Newton Police Department, which dispatches the private ambulance company that serves the city, also received reports indicating an explosion with injuries. Two ambulances and several police cars were dispatched to the scene.

The first companies to arrive reported smoke coming from the building but no evidence of a major fire or explosion damage. Engines 7 and 3, Ladder 2, and Car 2, the on-duty Assistant Chief, took positions in the parking area near the front entrance to the building and made contact with several employees who described the situation. There was no particular employee who appeared to be in charge or to be specifically responsible for providing information to the fire department. The shift supervisor had been injured in the initial explosion and was being treated by other employees.

The employees advised that the material on fire was sodium, that approximately 100 pounds of sodium was involved, and emphasized that it was important not to use water on it or to allow it come in contact with water. The employees also advised that salt should be used as the extinguishing agent. The firefighters were not advised that there could be water in the area from the previous wet washing operation.

Note: Sodium chloride is used as an extinguishing agent at this plant, because large quantities of pure dry sodium chloride are used for metal processing and the company keeps plenty on hand at all times. Large salt containers, designated for fighting combustible metals fires, are distributed around the plant. Scoops to apply the salt are kept in the containers, under a lid that is designed to exclude moisture.

These salt containers could be moved around on wheeled dollies and brought to the area where the material was burning. Additional quantities of bagged salt were stored in another area of the plant and could be used as a backup supply.

The assistant chief's aide, accompanied by a plant employee, made an initial reconnaissance entry to the fire area, which was approximately 200 feet back into the building. He was able to get close enough to open the doors to the fire room and see that there was a barrel burning and additional glowing material on the floor, apparently in the drip pan. There was a considerable amount of smoke in the interior; however, he was able to make this entry to the fire area and back without protective clothing or breathing apparatus. He returned to report his observations to the Assistant Chief, who had assumed command of the incident and was in contact with plant personnel inside the front door to the building. The Aide then went with a plant employee to open doors and activate exhaust fans to ventilate the interior of the building.

The assistant chief had been told that the drum contained about 100 lbs. of sodium, which would take all night to burn out. He was concerned about maintaining the integrity of the steel drum if the fire continued to burn and was told by an employee that the fire might burn through the bottom of the drum in three quarters of an hour. He was not aware of the fact that 100 pounds is far in excess of the normal amount of sodium that is burned in the disposal room or that the room had been used for wet washing earlier in the evening.

The assistant chief and the personnel from the first three companies were sufficiently familiar with the facility and confident in their previous experience with minor sodium fires that they felt they could handle the situation safely. They were familiar with the salt containers and enlisted the assistance of several employees to move some of the containers from other areas closer to the fire location. The assistant chief warned the crews to use all their standard protective clothing and self-contained breathing apparatus.

All three companies proceeded to the area of the disposal room. Several members worked with the plant personnel who were collecting salt containers from different parts of the building and staging them in the area outside the doorway to the fire area. The employees also obtained additional bagged salt from a storage area, but were directed to stay out of the immediate area of the fire by the firefighters.

The firefighters found several lumps of burning sodium in the drip pan and splattered around a small area, and a glowing red liquid in the drum, which they described as looking like volcanic lava. The liquid level in the drum was estimated at eight to ten inches deep. The drum was standing upright in the pan. Apparently the explosion had dislodged the metal strut and the weight at the bottom end of the drum had caused it to rotate to a vertical position. Heavy smoke was coming from the open top of the drum and visibility in the area was very limited.

Two lieutenants entered the small room to apply the salt, giving their first attention to the burning metal in the drip pan. The other firefighters scooped salt from the containers and passed the scoops to the lieutenants. A shovel that was found in the area was also used to bury the burning material in mounds of salt. Within a few minutes they had successfully buried all the exposed burning sodium and this part of the fire was considered under control, although the buried metal continued to flow red through the salt.

The two lieutenants then turned their attention to the fire in the drum. One grasped the edge of the drum with a gloved had and tilted it slightly to provide better access. Several scoops of salt were dumped into the drum by one of the officers with no apparent effect. The other lieutenant picked up the shovel and used it to transfer more salt into the burning drum.

When the first shovel of salt was dumped into the drum, there was a violent explosion. A fireball enveloped most of the area and burning molten sodium erupted up and out of the drum, splashed off the walls and ceilings, and splattered on the firefighters. All of the firefighters in the area were knocked off their feet and away from the center of the blast, bouncing off walls and equipment. Their leather helmets were blown from their heads as the hydrogen fireball passed through the area, then the molten sodium landed on them. Their distance from the drum determined how much of the burning metal landed on each individual.

The two lieutenants, who were in the room with the drum, were splashed with the molten metal virtually from head to toe. Their protective clothing and station uniforms were severely burned and both received critical, life threatening third-degree burns to major parts of their bodies. One of the lieutenants was able to remove himself from the room, but the second lieutenant had to be removed by rescuers.

Six firefighters and a third lieutenant had been involved in moving the salt containers and passing salt to the two lieutenants. They were all in the area outside the room, and were burned by the combination of the fireball and the spray of molten sodium. The assistant chief's aide was slightly burned by the fireball as he was returning to the fire area after completing his ventilation assignment.

The injured members extricated themselves and helped each other out of the immediate area. They had difficulty maneuvering in the tight space with the salt containers crowded into the narrow area near the room where the explosion occurred. The burning sodium that was splashed on their protective clothing continued to burn as they tried to exit toward the front of the building. Plant employees stopped them from leaving the building, warning that the moist outside air would make the sodium burn more intensely, and helped them remove their SCBAs and protective clothing in an interior corridor.

The plant employees poured mineral oil on the burning sodium to cool the material and the firefighters' burns. The assistant chief and two additional firefighters received burns to their hands attempting to help the injured firefighters remove their burning clothing. The molten metal was extremely hot and, where it landed on the protective clothing, it continued to penetrate through the layers to the skin. The sodium also reacted with moisture in the air and perspiration on the firefighters' skin to form corrosive sodium hydroxide.

The assistant chief, who was in the building's main corridor, was also knocked down by the force of the explosion. He made an urgent call for a second alarm and ten ambulances. Two units and a supervisor from the private ambulance company had been dispatched by the Newton Police Department on the original call and were already staged outside the complex when the explosion occurred. The additional units were dispatched when the Newton Fire Communications Center requested them through the police department. Following standard operating procedures for a hazardous materials incident, the ambulances and the supervisor had staged outside the immediate area, on Charlemont Street, anticipating that any patients would be brought out to a triage area in a safe room.

The burned members were all gathered in one area inside the building and were being treated by other firefighters and plant workers. Attempts to direct the ambulances to come to that location were unsuccessful for several minutes, because the ambulance radios were set with the Newton Police as the priority channel on their scan feature. The volume of police radio traffic prevented the messages from the Fire Department Incident Commander from getting through to the ambulances. This delayed the arrival of paramedics to treat the burned patients for several frantic minutes, although ambulances were staged only a block away.

Eleven fire department members were transported to area hospitals, including the burn centers at the Massachusetts General Hospital and at Brigham and Women's Hospital, both in the city of Boston. The presence of two burn centers in the metropolitan Boston area allowed the most seriously burned patients to be distributed to share the workload for treatment. Eight were admitted, two in critical condition, and three were treated and released with less severe burns. The two lieutenants were admitted in extremely critical condition; three months after the incident one remained in the Burn Center at Massachusetts General Hospital, still in very critical condition, with burns over 85 percent of his body.

FIRE CONTROL

The assistant chief called for a second alarm immediately after the explosion. Fire Chief Joseph Daniele, who had been monitoring the radio traffic at home, responded to the scene and requested a third alarm shortly after his arrival. The regional hazardous materials response team, which includes members from several suburban Boston fire departments, also responded. All firefighting operations were suspended until an assessment of the hazards and the risk of additional explosions could be made.

The sodium burned itself out within an hour and only a very small amount of residual fire was left for extinguishment. The fire was contained to the drum cleaning enclosure and smoke damage to the rest of the building was minor.

Massachusetts State Fire Marshal, F. James Kauffman, who was also monitoring the radio traffic, also responded to the scene and activated a major response of investigators from his division of the Department of Public Safety to assistant the Newton Fire Department

ANALYSIS

The immediate cause of the explosion was determined to be the introduction of water to the molten sodium. Physical evidence confirmed that the explosion occurred within the drum and did not involve the material that had been spilled or the water under the floor grates. The water is believed to have come from the shovel, which had been placed on the wet floor grate. The moisture was transferred to the salt as it was shoveled into the sodium drum.

Since sodium reacts violently with many substances, the contaminant may have been something other than water. Even rust on the shovel could cause a reaction; however, water is the most likely cause of the reaction.

The primary causal factor of this incident was the inappropriate actions of employees at the H. C. Starck facility, trying to burn a much larger heel of waste sodium than the normal operating procedure could accommodate. In addition, the design of the drum cleaning area, to include wet washing and sodium disposal in the same enclosure, created an unnecessary hazard. The requirement to keep sodium away from any potential source of moisture should have precluded conducting both of these operations in the same space.

The documented procedure for burning small amounts of waste sodium in the bottoms of drums appears to be reasonably safe, if all of the steps are followed. The appropriate procedure for dealing with a larger amount of residual sodium would be to maintain the argon in the empty space above the sodium, reseal the drum, and return it to the supplier for disposal or recycling. The lack of a documented procedure for dealing with larger quantities created a situation where employees had to make a judgment that could not be based on experience or an appropriate risk evaluation.

The company and the Newton Fire Department had done a good job of working together to familiarize firefighters with the facility and to inform them of the types of situations they might face at the plant. The training, however, was directed toward small sodium spill fires that can usually be handled with a relatively minor risk to personnel, as long as basic safety precautions are followed and standard operating procedures are employed. The dangers that were involved in this situation, with a drum containing molten sodium, were far more severe than the fire department or the company had anticipated. This situation shows how training that deals with low risk situations can create a false sense of security when the situation is more complex and dangerous.

If all of the information had been provided to the fire department, including the significance of the quantity of sodium that was involved, the fact that the sodium was in a molten state, and the presence of water in the area, a thorough risk analysis could have been considered, prior to formulating an action plan. The actual situation that was encountered was an extremely high risk hazardous materials incident. The risk evaluation would have revealed:

1. The presence of burning molten sodium in the drum was a different and much higher risk incident than previous training had anticipated.

2. There is no safe procedure available for a municipal fire department to deal with a molten sodium fire, unless it has been previously trained specifically for this type of situation and provided with specialized equipment and protective clothing. Structural protective clothing and self-contained breathing apparatus are inadequate protection for the risk of being splashed with molten sodium or any molten metal. There is no protective clothing designed to protect the user from direct contact with burning molten sodium. (A detailed analysis of the performance of the protective clothing and breathing apparatus are provided in the following section of this report.)

3. The possible extinguishment alternatives for molten sodium are very limited: blanketing the molten sodium with an inert gas or covering it with soda ash. In either case, the extinguishment agent, if available, would have to be applied very gently and carefully to float on top of the molten metal surface. It may also have been feasible to smother the fire with a lid that completely covered the opening, if one had been available. However, all of these possibilities would expose firefighters to excessive risk. The potential consequences of an error in the extinguishing procedure are extreme.

4. In the worst case scenario, if the container had failed, the molten sodium could have flowed through the grates and into the water containing troughs, resulting in an immediate explosion, probably of equal or greater magnitude to the explosion that did occur. The fire was burning in an isolated room, designed to contain an explosion. If no one had been in the immediate area when the explosion occurred, there would have been no injuries and the damage would have been about the same as actually occurred, which was minor. If the doors to the room had been closed even less damage would have occurred.

Weighing the alternatives that were available to the Newton Fire Department, a thorough risk evaluation would suggest taking no action on this fire. The risks involved in taking action are very high, while the consequences of not fighting the fire are relatively minor. The fire probably would have continued to burn, producing smoke, until the material was fully consumed. In all probability, the steel drum would not have failed and the fire would have terminated when the fuel was consumed.

The post incident evaluation suggests that a decision to not fight the fire would have been the best decision. However, it was very difficult to recognize these factors at the time, based on the information that was available to the Incident Commander and the other firefighters. Their training had not prepared them for the situation they encountered, although they were much better informed and trained than most municipal firefighters on dealing with sodium fires. They had been trained to handle a less hazardous situation and had not been trained or given the information that would have caused them to recognize the higher risk situation.

The failure to establish an effective liaison with knowledgeable plant employees caused decisions to be made before all of the information could be gathered, assembled, and analyzed. Instead of one primary liaison contact between the fire department and a responsible individual for the plant, several different fire department members had contact with several different plant employees, which increased the level of confusion.

The interviews with injured fire department members indicate that the crews did not have a good appreciation of the nature of the situation before they went in to attempt extinguishment. The smoke filled atmosphere made it difficult to size-up the situation; however, they were guided by the plant workers' confirmation that they should use salt as an extinguishing agent. They did not have a more specific plan of action to establish what they would do after entering the fire area.

Less than ten minutes elapsed from the arrival of the first companies until the explosion occurred. This suggests that very little time was taken for evaluation of the situation and formulation of a plan. This indicates that the approach was very action oriented, when the actual situation should have called for careful analysis before, or instead of, taking action.

The delay that occurred in contacting the EMS units to come in and treat the injured firefighters indicated a deficiency in an otherwise good standard operating procedure. Having EMS units respond and stage at a safe distance from the incident is a good plan for hazardous materials incidents, particularly when the EMS providers are not trained or equipped to operate in a dangerous area. The weakness in the plan was the inability to make contact between the Incident Commander and the EMS supervisor when the ambulances were needed at a specific location. The delay of several minutes caused an extremely high level of anxiety, although it does not appear to have had significant consequences on the outcome of the incident; the paramedics did not have any better treatment capability for the burned firefighters than was already being attempted and faster transportation would have made little difference in the outcome.

Note: The communications problem between the Fire Incident Commander and the ambulances has already been corrected by changing the standard operating procedure. The ambulance crews had been provided with an additional radio to maintain contact with the Fire Incident Commander.

PROTECTIVE CLOTHING

The protective clothing worn by the personnel involved in this situation was designed for structural firefighting and could not be expected to protect the user from exposure to a molten metal fire. It is significant to note the performance of the protective clothing and equipment and the protection that it did provide, recognizing that there is no standard for protective clothing for the actual situation that was encountered.

All of the injured personnel, except the assistant chief's aide, were wearing a basic ensemble consisting of black Nomex® turnout coats, leather helmets, leather gloves, 3/4 length rubber boots, and self-contained breathing apparatus. The aide was wearing a station uniform without protective clothing or SCBA.

Turnout Coats – Most of the turnout coats were several years old and appear to have been purchased prior to the adoption of the NFPA standards for protective clothing. Only one coat, worn by a younger firefighter, had a label indicating compliance with NFPA Standard 1971. Most of the coats did not have labels to identify the constituent materials or construction standards and appear to date back to the 1970s. The outer shells are black Nomex®, approximately 7.5 oz. per square yard, some impregnated with neoprene to make the outer layer impermeable. (The materials appear to be generally acceptable with respect to the NFPA standards that were first adopted in 1975.)

Most of the liners appear to be needle punch Nomex® with neoprene moisture barriers and several had webbing sections in the armpit areas to allow for ventilation. One coat had a quilted liner, with no moisture barrier, but was worn under an outer shell that incorporated a neoprene moisture barrier in the shell. Another coat was worn with a nylon winter liner, in addition to the regular liner.

One coat was worn with no inner liner; only the outer shell provided protection to the user. Wearing a structural fire suppression coat in this manner is a dangerous practice, as much of the thermal protection provided by a turnout coat depends on the insulation provided by the liner.

The damage to the coats indicates that they withstood the initial fire ball created by the hydrogen explosion, without significant damage or failure. This fire ball probably lasted for only a second (or a fraction of a second) and the turnout coats would be expected to provide reasonably good protection to the body from this type of exposure. Burns could be anticipated anywhere the protection of the coat was compromised, such as an unsecured collar closure. The coat that was worn without a liner could be inadequate to prevent burns from this type of exposure.

Where the coats were splashed with molten sodium, it burned or melted through the outer shell and continued to penetrate into the liner material. The degree of damage indicated the pattern of molten sodium impact on the individual coats. The two lieutenants who were in the room with the sodium drum were directly splashed by the molten metal, which burned large sections of the coats and liners and penetrated through to the station uniforms, resulting in large areas of third degree burns to their bodies. In each case the major burn areas coincided with their orientation toward the drum.

Most of the other members were splashed or splattered by varying amounts of molten sodium, depending on their distance from the drum. The splatter patterns were visible where the sodium burned through the coats and into the liners. In most cases the sodium did not burn all the way through the liners; however, the heat of the burning molten sodium would be sufficient to cause second or third degree burns to the skin in these areas.

Boots – All of the personnel wore 3/4 length rubber boots, some in the extended position and some in the folded down position, as noted from the damage to the boots. This is believed to have contributed to some leg and thigh burns on the firefighters who had their boots turned down. There was no evidence of molten or burning sodium on the boots; however, the exposed surfaces were coated with sodium oxide.

Helmets – The helmets were all leather and also appeared to be several years old. Due to the absence of labels it was impossible to determine their vintage or specific design features; however, none appeared to meet any of the newer editions of NFPA Standard 1974. Most of the helmets had been retrofitted with chin straps, but many were found to be secured around the brim of the helmet instead of being used to hold the helmets in place. All of the helmets are believed to have been blown off the wearers' heads by the explosion, exposing their heads and the retaining system of their SCBA face masks to the fire.

The helmets were also fitted with Bourke eye shields, but all of the eye shields appeared to be in the flip-up position, indicating they were not in use when the explosion occurred. It was impossible to determine if the ear flaps had been in use at the time of the explosion, because the liners and ear flaps of several helmets had burned away.

Gloves – The gloves that were examined were leather and appeared to comply with current or recent editions of NFPA Standard 1973. The gloves protected the hands of the members who were wearing them. At least one member had removed a glove and suffered serious burns to that hand. Some wrist burns were reported where the knit wristlets bunched-up and allowed some penetration of molten metal.

Station Uniforms – Most of the personnel were wearing non-fire retardant station uniforms under their protective clothing. The Newton Fire Department was in the process of issuing fire retardant (FR) station uniforms, complying with NFPA Standard 1975; however, they were being phased-in and only one of the injured personnel is believed to have been wearing an FR uniform. Most of the personnel still wore the previous issue uniform items, which were cotton/polyester blends. (One label indicated a 65 percent polyester content and 35 percent cotton content on a particular item.) Some also wore items of non-uniform clothing.

The objective of the fire retardant station uniform standard is to avoid materials that could contribute to a burn injury by igniting or melting. The one FR uniform that was worn in this incident showed no evidence of damage, while most of the non-FR items showed some damage where they were most directly exposed to the molten sodium. The non-FR station uniforms of the two most seriously burned members were seriously damaged.

It is doubtful if the FR clothing would have made a significant difference in the burn injuries to the members who were splashed directly by molten sodium. In the cases where burning sodium splattered on turnout coats and burned into the liners, the burns might have been reduced if the members had been wearing FR uniforms. The most significant area of exposure for the station uniforms was the upper legs, thighs, and buttocks, where the uniform pants provided the only protection, particularly where the members had their boots folded down.

Evidence of the heat content of the molten metal could be seen in a pair of eyeglasses that were found in the shirt pocket of one of the firefighters. Molten sodium had burned through the coat and liner, penetrated the outer layer of the shirt material, and then penetrated through the lens of the eyeglasses, leaving a metallic luster on the inner and outer surfaces.

Breathing Apparatus – The self-contained breathing apparatus included both Scott 4.5 and Scott 2A units, depending on the company to which the individual was assigned. The shoulder and waist straps on some of the 2A units melted from heat exposure and released from the wearers' bodies. (The straps had not been upgraded from the original black webbed material that has been noted in previous incidents to be susceptible to very rapid melting when exposed to flames.) There was also damage to the low pressure breathing tubes on some of these units, including one which separated at the regulator fitting. The face masks on the 2A models stayed in-place on the users heads, since the heavy rubber "spider" straps resisted the heat exposure.

The Scott 4.5 units have fire resistant shoulder and waist straps and no problems were noted with the main harnesses; however, the facepiece retaining systems on these units were compromised by heat exposure. The netting and the single take-up straps that hold the facepiece on the user's head were exposed to the fireball when the helmets were blown off and the straps and/or the netting failed on all of the 4.5 units. This released the facepiece from the user's head. On one unit the lens also separated from the rubber body of the mask along the upper part of the face piece. Some of the burns to the face, neck, and ears were particularly severe.

The facepieces of the members who were close to the explosion were completely coated with a mixture of sodium and sodium oxide which made them opaque and impossible to see through. The members trying to escape from the fire had to remove their facepieces to see anything. Where molten sodium landed on the lenses, it penetrated through the plastic material leaving a metallic coating on the inside and outside. Also, on one of the 4.5 SCBAs that was splashed directly with molten sodium, the facepiece mounted regulator was completely destroyed, including the metal parts.

All of the SCBAs were equipped with Personal Alert Safety System (PASS) devices; however, they did not appear to have been turned on at the time of the explosion. The PASS devices that were examined were still functional, but showed signs of heat exposure while in the off position. No one reported using or hearing an alarm from a PASS device.

Comment – Protective clothing and breathing apparatus that is designed for structural fire suppression does not provide adequate protection for exposure to molten metals. Molten metals, particularly burning molten metals, present a much more severe danger than ordinary structure fires. Only special protective clothing, designed to resist very high temperatures and to shed molten metals, should be used where there is a potential of coming in any contact with molten metals. There is no practical protective ensemble that would offer adequate protection for an individual splashed by burning molten sodium. Firefighters should avoid situations where there is a danger of being splashed by any molten metal.

The Newton Fire Department was in the process of evaluating newer protective ensembles, which incorporate protective trousers (bunker pants) instead of 3/4 length rubber boots. This level of protection is required by current NFPA standards; however, the acceptance of the concept has been slow in some geographic areas, particularly in New England. The analysis of the injuries in this incident clearly indicates that the burn injuries would have been less extensive if the personnel had been wearing ensembles that meet the current editions of the NFPA standards.

The currently accepted level of protective clothing for structural fire suppression includes protective coats and pants, foot protection, gloves, helmets with chin straps, full ear and neck protection, fire resistant station uniforms, self-contained breathing apparatus, and PASS devices. Most of the items used by the members involved in this incident did not met current design standards or were used improperly, or both. A full protective ensemble, meeting the standards for structural fire suppression, could prove adequate for an incident involving a small amount of burning molten metal, where the risk of coming in contact with molten metal is remote. This level of protection should be limited to situations where the greatest risk would be to come in contact with a few droplets of molten metal, which would not penetrate the outer shell of protective coat and pants. In this type of situation a face shield should be used over the SCBA facemask for extra protection from metal splatter on the facemask lens.

In a situation where a significant amount of molten sodium was involved and splashed directly on the firefighters, even a full set of modern structural protective clothing would not have been adequate to prevent injury. However, the degree of injury could have been reduced if the clothing had met current standards or if all the clothing that was available had been worn properly. The only way to avoid injury would have been to avoid the risk by taking no action to extinguish the fire.

CRITICAL INCIDENT STRESS

One of the most severe consequences of this incident was severe post incident stress, which particularly affected many of the individuals who were involved in assisting the injured firefighters. The victims could not help themselves and the rescuers were frustrated in their attempts to help them.

All of the personnel in the fire area were burned and some were in very critical condition. The sodium that had impregnated their protective clothing continued to burn the injured members as others tried to help them. The most common burn treatment, applying cool water, would only have made their injuries more severe.

Additional stress was created by the realization that the injured members had not been adequately protected by their protective clothing. Personnel who had not previously encountered molten metal burns did not appreciate the severity of the burns or the inadequacy of structural protective clothing to prevent them.

Immediate counseling assistance was provided by the regional traumatic incident response team and several individuals have been referred for further treatment.

LESSONS LEARNED

1. **The extreme hazard of molten sodium must be recognized.**

    Molten sodium is an extremely hazardous material, reactive to water and most other extinguishing agents. Previous experience, pre-fire planning, and information provided by plant employees caused firefighters to believe they could handle the burning sodium without excessive risk.

2. **The best action plan would have been No Action.**

    A full evaluation of the risks and potential consequences of this incident, after the fact, leads to the conclusion that the safe plan would have been to take no action.

3. **Structural protective clothing is not designed for molten metal.**

    Structural protective clothing and self-contained breathing apparatus are not designed to provide adequate protection for exposure to molten metal.

4. **The way protective clothing was worn decreased its effectiveness.**

    Members involved in the incident would have been considerably better protected if they had been wearing full protective clothing ensembles that meet current standards, including protective trousers (turnout pants) instead of 3/4 length boots. They would have had better protection if they had used the chin straps on their helmets, pulled up 3/4 length boots, and (in one case) not worn a turnout coat without the liner.

5. **Extinguishing agent choice has limitations.**

Sodium chloride is effective as an extinguishing agent for small quantities of burning sodium, but it is ineffective on molten sodium. Sodium chloride may be dangerous if it is not completely dry. Other agents offer higher degrees of safety for small sodium fires.

6. **The number of injuries reflects the fact that the hazard was not recognized.**

Because the hazard was not recognized, three companies of firefighters were in the immediate area assisting in the extinguishment efforts. All of the personnel in the area were burned when the explosion occurred.

AFTERWORD

During the start-up procedure at the H. C. Starck facility, after the process had been shut down for investigation of the explosions, another fire occurred. The system had been shut down for more than two months and as it was being reheated a minor sodium leak occurred. The leaking sodium ignited and the area was evacuated. The Newton Fire Department responded and isolated the area, but took no action. The system was shut down and the fire was allowed to burn itself out. There was no significant damage from this incident and there were no injuries.

The company was cited by the U.S. Department of Labor for OSHA violations as a result of this incident. The violations related to the burning of sodium in an area where molten sodium and water could mix, and the company was required to have a person qualified in process design approve any deviations from standard operating procedure when burning excess sodium.

The company has changed its policy and now returns all used drums to the supplier for disposal. Plant personnel are being trained to function as an individual fire brigade.

APPENDIX A

Sodium MSDS Sheets and Other Materials on Sodium Properties, Uses, Storage, and Handling

Appendix A (continued)

Du Pont Chemicals

1160CR	Revised 03-Sep-93	Printed 07-Sep93

S o d i u m

MATERIAL IDENTIFICATION

Corporate Number	DU001251
Manufacturer/Distributor	DuPont 1007 Market Street Wilmington, DE 19898
Phone Numbers	Product Information I-800-441 -9442 Transport Emergency CHEMTREC: I-800-424-9300 Medical Emergency I-800-441-3637
Grade	REGULAR (STANDARD); NIAPURE (LOW CALCIUM)
Chemical Family	ALKALI METAL
Trade Names and Synonyms	SODIUM METAL "NIAPURE" SODIUM "NIAPURE" is a non-registered trademark of Du Pont.
CAS Name	SODIUM
CAS Number	7440-23-5
Formula	Na
TSCA Inventory Status	Reported/Included
NFPA Ratings	Health: 3 Flammability: 1 Reactivity: 2 Water Reactive
NPCA-HMIS Ratings	Health: 3 Flammability: 1 Reactivity: 2 Personal Protection rating to be supplied by user depending on use conditions.

(continued)

Appendix A (continued)

COMPONENTS

Material	CAS Number	Percent
SODIUM	7440-23-5	100

REACTION PRODUCT WITH WATER:		
SODIUM HYDROXIDE	1310-73-2	

PHYSICAL DATA

Boiling Point	881°C (1,618°F) at 760 mm Hg.
Vapor Pressure	1 mm Hg at 493°C (920°F)
Vapor Density	Not applicable
Melting Point	97.8C (208°F)
Evaporation Rate	Not applicable
Water Solubility	Reacts violently with water.
Odor	Odorless
Form -	Metallic solid
Specific Gravity	0.97 at 20°C (68°F)

pH Information : Reacts with water to form sodium hydroxide (high pH) and hydrogen gas.

Color : In inert atmosphere- pinkish silvery when fresh cut.
In air- white to gray.

HAZARDOUS ACTIVITY

Instability	Stable.
Decomposition	Decomposition will not occur.
Polymerization	Polymerization will not occur.

Incompatibility : Reacts violently with any materials containing water and many materials containing oxygen, halides, or active hydrogen. Reaction with water gives sodium hydroxide and hydrogen gas, which may explode. Burning produces sodium oxide fumes.

(continued)

Appendix A (continued)

FIRE AND EXPLOSION DATA

Flash Point	Not applicable

Flammable Limits in Air, % by Volume LEL Not determined
                                     UEL Not determined

Autodecomposition	Not applicable

Autoignition : -120-125 deg C (-248-257 deg F)

The Autoignition Temperature range varies widely depending on pool or droplet size, air temperature, velocity, humidity, etc.

Fire and Explosion Hazards

Flammable solid.

Reacts violently with water releasing hydrogen gas, which will ignite and explode in air. Burning produces dense, white, irritating smoke. Follow appropriate National Fire Protection Association (NFPA) codes.

Extinguishing Media

Dry Soda Ash, Light (low density, floats on molten sodium) or Class D fire extinguisher. Dry salt or sand is less effective, but can be used.

Special Fire Fighting Instructions

DO NOT use water. Do not use Carbon Dioxide (CO2), soda-acid, or chlorinated fire extinguishing agents such as carbon tetrachloride. Stay upwind and use self-contained breathing apparatus if needed. Sodium melts and burns with little or no flame, but with yellow to yellow-orange glowing globules that appear to move on the surface of the molten pool. Reduce fire by diking to limit sodium surface, then smother with soda ash or cover with a steel lid.

HEALTH   HAZARD INFORMATION

Causes severe eye and skin burns from reactions to sodium hydroxide-- effects may be permanent. Fumes from sodium reactions to sodium oxide may irritate the nose, throat, and lungs. Ingestion will cause burns of the gastrointestinal tract with perforations by formation of sodium hydroxide.

Sodium reacts rapidly with 'moisture in air or tissues to form sodium hydroxide and sodium oxide. Effects following inhalation, ingestion, or skin or eye contact result from direct chemical reaction with tissue and from thermal reaction with water.

ANIMAL DATA:

   Oral ALD: 500 mg/kg in rabbits (sodium hydroxide)

Sodium is very corrosive to animal skin and eyes by reactive' formation of sodium hydroxide. Toxic effects described in animals from exposure by inhalation or ingestion include irritation of the respiratory tract, and extensive necrosis of the gastrointestinal tract.

(continued)

Appendix A (continued)

HEALTH HAZARD INFORMATION (continued)

HUMAN HEALTH EFFECTS:

Overexposure by skin or eye contact include skin burns or ulceration; or eye corrosion with corneal or conjunctival ulceration. By inhalation, the effects include irritation of the upper respiratory passages with coughing and discomfort. By ingestion, the effects include abdominal discomfort characterized by nausea, severe pain, diarrhea, and collapse.

Carcinogenicity

None of the components in this material is listed by IARC, NTP, OSHA, or ACGIH as a carcinogen.

Exposure Limits

TLV (ACGIH)
PEL (OSHA)

None Established
Particulates Not Otherwise Regulated
15 mg/m3 - 8 Hr TWA - Total Dust
5 mg/m3 - 8 Hr TWA - Respirable Dust

Other Applicable Exposure Limits
SODIUM HYDROXIDE
AEL* (Du Pont)
TLV (ACGIH)
PEL (OSHA)

2 mg/m3 - 15 Min. TWA
2 mg/Jm3 (Ceiling)
2 mg/m3,8 Hr TWA

Du Ponrs Acceptable Exposure Limit. Where governmentally imposed occupational exposure limits which are lower than the AEL are in effect, such limits shall take precedence.

Safety Precautions

Persons handling sodium should be thoroughly familiar with its hazards and proper first aid procedures. Do not get in eyes, on skin, or on clothing, and avoid any contact with water. Avoid breathing fumes from sodium reactions.

FIRST AID

INHALATION
If fumes from sodium reactions are inhaled, remove to fresh air immediately. If not breathing, give artificial respiration. If breathing is difficult, give oxygen. Call a physician.

SKIN CONTACT
In case of contact, immediately remove particles of sodium adhering to the body or clothing. This must be done before washing to avoid additional heat from the reaction between sodium and water. After the particles are removed, flush the skin with plenty of water for at least 15 minutes while removing contaminated clothing and shoes. Call a physician. Burn or wash clothing and shoes.

EYE CONTACT
In case of contact, remove sodium, then immediately flush eyes with plenty of water for at least 15 minutes. Call a physician.

INGESTION
If swallowed, do not induce vomiting. Give large quantities of water. Call a physician immediately. Never give anything by mouth to an unconscious person.

(continued)

Appendix A (continued)

PROTECTION INFORMATION

Generally Applicable Control Measures and Precautions

Good general ventilation should be provided to keep fume concentrations below the exposure limits and to prevent the accumulation of hydrogen gas.

Personal Protective Equipment

Solid Sodium: Sodium bricks can be handled safely using chemical splash goggles and DRY moleskin mitts. Mitts should be oversized for easy removal and should extend up the arms to prevent caustic burns. Wear a long sleeve shirt. A full-length face shield, alkali resistant apron, and other special protective equipment may be needed for specific jobs.

Molten Sodium: When liquid sodium is handled or there is danger of spillage, full protective flameproof clothing should be available and used as appropriate. This includes: hard hat with brim: chemical splash goggles: full length face shield: fire resistant ("Nomex" Aramid Fiber or alternative) long underwear, pants and shirt (or coveralls) neck shroud, spats, apron: and heavy duty work shoes. Two, or preferably more, layers of flameproof d othing are clearly more effective than one layer. A NIOSH/MSHA air supplied or self-contained breathing apparatus.is needed if large amounts of sodium oxide smoke are present: a NIOSH/MSHA approved air purifying respirator can be used for smaller amounts.

DISPOSAL INFORMATION

Aquatic Toxicity

Sodium Hydroxide (reaction product with water)
48-hour TLm, bluegill sunfish: 99 mg/L
96-hour TLm, mosquito fish : 125 mg/L

Spill, Leak, or Release

NOTE: Review FIRE AND EXPLOSION HAZARDS and SAFETY PRECAUTIONS before proceeding with clean up. Use appropriate PERSONAL PROTECTIVE EQUIPMENT during clean up.

Cover with DRY Soda Ash, Light; shovel into a dry metal container and cover again with soda ash, and dispose of promptly. Avoid putting wet sodium in a covered container because a hydrogen ex plosion may occur. Wear proper protective equipment. Comply with Federal, State, and local regulations on reporting releases. The CERCLA Reportable Quantity is 10 lbs.

Waste Disposal

Comply with Federal, State, and local regulations. If approved, may be burned in an incinerator equipped with a scrubber. Small amounts of sodium can be disposed of by weathering, by steaming (which requires special instructions), or by burning in open air, if approved. Considerable white smoke will develop when burning even small amounts of sodium. Sodium disposal, and disposal of empty drums, should not be attempted by inexperienced people. Contact DuPont for technical information or call a lcensed disposal contractor. This material may be an RCRA Hazardous Waste upon disposal due to the reactivity characteristic.

(continued)

Appendix A (continued)

SHIPPING INFORMATION

DOT/IMO Proper Shipping Name	SODIUM
Hazard Class	4.3
UN No.	1428
DOT/IMO Label	DANGEROUS WHEN WET
Packaging Group	II
Shipping Containers	Tank Car Tank Truck Drums Samples: Fused, in 1 quart tin cans: 2 1/2 lb. bricks in 5 gallon pails Reportable Quantity : 10 lbs/4.54 kg

STORAGE CONDITIONS

Store in segregated area of fire resistant, watertight building without sprinklers, steam, water lines, skylights, or potential for flooding. Ventilate to avoid hydrogen accumulation. Keep drums covered to prevent caustic formation from moisture in air. Keep from possible contact with water. Nitrogen purging of open drums will minimize reactions with moisture and oxygen in air. Keep drums tightly closed. Do not store with combustibles or flammables as firefighting problems would be compounded. Use only clean, dry utensils in handling.

WARNING:

Bulged drum indicates hydrogen gas pressurization. Puncture drum from at least six feet away to vent pressure before removing lid.

TITLE III HAZARD CLASSIFICATIONS

Acute	Yes
Chronic	No
Fire	Yes
Reactivity	Yes
Pressure	No

LISTS:

SARA Extremely Hazardous Substance   -NO
CERCLA Hazardous Material   -Yes
SARA Toxic Chemical   -NO

CANADIAN WHMIS CLASSIFICATIONS:

B6; E

(continued)

Appendix A (continued)

ADDITIONAL INFORMATION AND REFERENCES

For further information, see DuPont "Properties, Uses, Storage, and Handling" Bulletin.

The data in this Material Safety Data Sheet relates only to the specific material designated herein and does not relate to use in combination with any other material or in any process.

Responsibility for MSDS: Dupont Chemicals
Engineering & Product Safety
P. O. Box 80709, Chestnut Run
Wilmington, DE 19880-0709
302-999-4946

# Indicates updated section.

End of MSDS

Appendix A (continued)

SODIUM.

Properties, Uses, Storage and Handling

Appendix A (continued)

IMPORTANT

SAFETY

INFORMATION

CAUTION: KEEP SODIUM AWAY FROM WATER. SODIUM METAL CAUSES SEVERE BURNS TO SKIN AND EYES. FUMES FROM SODIUM REACTIONS MAY IRRITATE NOSE, THROAT AND LUNGS IF INHALED. SEE PERSONAL SAFETY AND FIRST AID SECTION.

TRANSPORTATION EMERGENCIES

IF SODIUM IS INVOLVED IN AN ACCIDENT OR EMERGENCY, CALL THESE NUMBERS FOR INFORMATION AND ASSISTANCE:

IN THE UNITED STATES

CHEMTREC I-(800) 424-9300
DUPONT I-(716) 278-5158

IN CANADA

CANUTEC I-(613) 996-6666
DUPONT I-(61 3) 348-3616

Appendix A (continued)

TABLE OF CONTENTS

Appendix A (continued)

Niagara's ample hydroelectric power is committed to industry by long-term contract.

New transmission facilities route power to Du Pont's sodium shop.

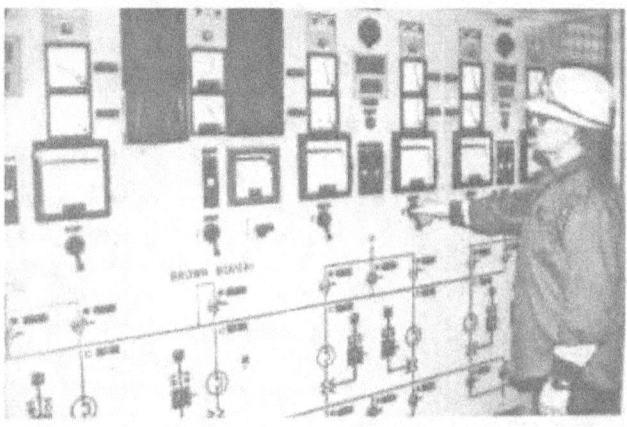

Modern distribution systems deliver reliable DC power to sodium cells.

2

Appendix A (continued)

Sodium has been manufactured at Niagara Falls, N.Y continuously since 1896 because of the readily available hydroelectric power. Du Pont has supplied approximately five billion pounds of product to sodium users through 1986, and has pioneered both the technology and handling procedures used in the industry today. Sodium technology developed at Niagara has been licensed worldwide. While sodium is an extremely reactive substance, Du Pont experience, readily shared with diverse users, has enabled many years of safe operation with this product. Du Pont's safety and handling expertise is available to all sodium users, both through this bulletin and by calling your Du Pont representative.

Sodium (Na, CAS Registry No. 7440-23-5) is a soft, low density metal easily cut with a knife. It is so reactive it does not occur as an element in nature, but is widely distributed in stable compounds. It was first isolated as a pure element in 1807 in England, and is manufactured today by electrolysis of sodium chloride in a high temperature bath of molten metal chlorides.

In an inert atmosphere freshly cut sodium has a pinkish, bright metallic luster. In air, the cut surface quickly forms a white or light grey coating of sodium oxide, hydroxide, and carbonate. Nitrogen, argon, and helium are inert to sodium and can be used to pad sodium containers. Simple hydrocarbons such as mineral oil, kerosene, xylene, and toluene neither dissolve sodium nor react with it, and are used as media for sodium dispersions. Sodium can be stored in high boiling hydrocarbons such as mineral oil or kerosene to retard oxidation. Sodium metal is a good conductor of heat and electricity. It is thermally stable in dry, inert atmospheres such as nitrogen or argon and will not explode or detonate from heat or shock.

Sodium is a powerful reducing agent: It reacts vigorously with water forming sodium hydroxide and hydrogen gas with release of considerable heat. Since the hydrogen usually ignites with explosive violence in the presence of air or oxygen, it is essential to avoid all water contact in sodium storage and handling. Sodium will auto-ignite in air at temperatures of approximately 120-125 C and above. Contact between sodium and halogenated hydrocarbons may cause an explosive reaction.

Du Pont offers sodium in regular and low-calcium grades.

SPECIFICATIONS AND TYPICAL ANALYSES

	Specifications	Analyses[a]
Sodium, wt. % (as produced)		99.9 +
Potassium, ppm		350
Calcium, ppm (regular grade)	400 max.	300
Calcium, ppm (NiaPure*)	10 max.	5
Chlorides (as Cl), ppm	50 max.	30
Boron, ppm		1

PHYSICAL PROPERTIES[b]

Atomic weight	22.99
Melting point, C	97.8
F	208.0
Boiling point, C	881.4
F	1618
Heat of fusion, cal/g	27.0
kJ/kg	113.0
Volume increase on melting, %	2.6
Heat of vaporization (at bp), cal/g	926
MJ/kg	3.87

	Solid		Liquid	
	20°C (68°F)	mp	mp	bp
Density, g/cm³				
(Mg/m³)	0.968	0.951	0.927	0.740
approx. lb/cu ft	60.4	59.3	57.9	46.2
approx. lb/gal	8.05	7.91	7.71	6.16
Viscosity, cp				
(mPa/s)			0.71	0.15
Surface tension,				
dyne/cm (mN/m)			192	113
Thermal conductivity,				
W/m K	132.3	119.3	87.0	48.6
Electrical resistivity,				
μΩ/m	469	660	964	5300
Specific heat,				
cal/g C	0.29	0.31	0.331	0.307
kJ/kg K	1.21	1.3	1.38	1.28

a. The above table gives typical analyses based on historical production performance. Du Pont does not make any express or implied warranty that future production will demonstrate or continue to possess these typical analyses. A full typical trace element analysis is available in Du Pont Bulletin—Sodium Uses And Advantages.

b. "Sodium-NaK Engineering Handbook," Vol. 1, O.J. Foust, Gordon and Breach, Science Publishers, Inc. (1972).

Low-calcium sodium (NiaPure*) typically contains less than 10 ppm calcium, and is available in limited quantities for critical heat transfer and other special applications. Large quantities of NiaPure* can be supplied with sufficient notice.

CHEMICAL PROPERTIES

Sodium melts at 97.8 C (208 F) to a silvery white, low-viscosity liquid resembling mercury. At temperatures around 120-125 C (248-257 F), liquid sodium ignites and burns in air with a yellow flame, giving off dense, acrid white smoke that is predominantly sodium monoxide (Na2O). Some sodium peroxide (Na2O,) also forms, especially at higher temperatures and in the presence of excess oxygen. If moisture is present, the oxides are converted to sodium hydroxide.

Trademark applied for by Du Pont Company.

3

Appendix A (continued)

Water reacts vigorously with either solid or liquid sodium in an exothermic reaction which liberates hydrogen gas. Under most circumstances when air or oxygen is present, explosions will accompany hydrogen gas generation.

CAUTION: Reaction rates in the examples cited below can vary widely with sodium particle size and temperature. The smaller the sodium particles or the higher the prevailing temperature, the more rapid the reaction. See NFPA's "Manual of Hazardous Chemical Reactions"[c] for a compilation of sodium reactions.

Sodium and hydrogen do not combine at room temperature. Temperatures of 200-350 C (390-660 F) are necessary for the formation of sodium hydride (NaH). At these temperatures dry ammonia gas also reacts with sodium to yield sodamide ($NaNH_2$). At 370-425 C sodium reacts with anhydrous sodium hydroxide to produce sodium monoxide and hydrogen:

$$Na + NaOH \rightleftharpoons Na_2O + \tfrac{1}{2}H_2 \qquad (1)$$

Low partial pressure of hydrogen over the system shifts the equilibrium to the right; high partial pressure of hydrogen shifts it to the left. With additional sodium, sodium hydride is also produced.

Sodium reacts with halogens to form the corresponding sodium halide. The reaction can be vigorous, depending on the halogen (F, Cl, Br) and the physical form of the metal. In an inert atmosphere, sodium reduces metal halides, e.g.:

$$TiCl_4 + 4\,Na \longrightarrow Ti + 4\,NaCl \qquad (2)$$
$$KCl + Na \rightleftharpoons K + NaCl \qquad (3)$$

Sodium also reduces potassium hydroxide and potassium cyanide in equilibrium reactions:

$$Na + KOH \rightleftharpoons NaOH + K \qquad (4)$$
$$Na + KCN \rightleftharpoons NaCN + K \qquad (5)$$

Phosphorus chlorides (PCl_3 and PCl_5) react with sodium, sometimes with detonation, to form sodium chloride and sodium phosphide (Na_3P). Phosphorus oxychloride ($POCl_3$) reacts explosively with sodium when heated.

Dry hydrogen sulfide gas and sodium combine to form sodium sulfide. In the presence of moisture, the reaction is rapid and evolves enough heat to melt the sodium. Carbon disulfide and sodium react violently, forming sodium sulfide, some disodium acetylide, and carbon.

Sodium and sulfur dioxide can be combined under controlled conditions to form sodium hydrosulfite ($Na_2S_2O_4$). The reaction is violent when the sodium is near its melting point.

Carbon dioxide gas shows little tendency to react with sodium at room temperature. At somewhat higher temperatures, sodium carbonate or, under controlled conditions, sodium oxalate is formed. **Carbon dioxide (CO_2)**

gas can support the combustion of sodium. Accordingly, neither CO_2 nor CO_2 powered fire extinguishers should be used on sodium fires. Solid CO_2 can react explosively with sodium on contact.

Alcohols react with sodium to form the corresponding alkoxide and hydrogen. Primary alcohols react more rapidly then secondary or tertiary alcohols. Reactivity declines as the molecular weight of the alcohol increases. The reaction affords a convenient source of hydrogen for controlled hydrogenations.

Naphthalene can be hydrogenated to 1,4-dihydronaphthalene with sodium in ethanol. Naphthalene with sodium in isoamyl alcohol yields 1,2,3,4-tetrahydronaphthalene.

In the Bouveault-Blanc reaction, esters are reduced by the reaction of sodium and an alcohol:

$$RCOOR' + 4\,Na + 2\,C_2H_5OH \longrightarrow RCH_2ONa + R'ONa + 2\,C_2H_5ONa \qquad (6)$$

Sodium alkoxides catalyze condensation reactions of the acetoacetic ester type:

$$2\,CH_3COOC_2H_5 \xrightarrow{C_2H_5ONa} CH_3COCH_2COOC_2H_5 + C_2H_5OH \qquad (7)$$

Alkyl and aryl halides are condensed by the action of sodium in the Wurtz-Fittig reaction:

$$2\,RX + 2\,Na \longrightarrow R\text{-}R + 2\,NaX \qquad (8)$$

Aliphatic dihalogen compounds may undergo cyclization in the Freund reaction:

$$XCH_2CH_2CH_2X + 2\,Na \rightarrow H_2C\underset{CH_2}{\overset{}{\diagdown\diagup}}CH_2 + 2\,NaX \qquad (9)$$

Tetraethyl lead is produced by reacting an alloy of sodium and lead with ethyl chloride:

$$4NaPb + 4C_2H_5Cl \longrightarrow (C_2H_5)_4Pb + 3Pb + 4NaCl \qquad (10)$$

Chloromethanes (CCl_4, $CHCl_3$, etc.) dehalogenate explosively with sodium. **(Since presently known halohydrocarbon fire-extinguishing products are either hazardous or ineffective on sodium fires, they should be excluded from areas where sodium is handled or stored.)**

Finely divided sodium and cyclopentadiene in tetrahydrofurand form cyclopentadienylsodium. Molten sodium reacts with acetylene to form sodium acetylide, which is used to introduce the triple bond into other molecules. Sodium dispersions react similarly with acetylenes.

National Fire Protection Association, 470 Atlantic Ave., Boston, MA 02210

dAlso a Du Pont product

Appendix A (continued)

Du Pont has manufactured sodium in Downs-type cells since 1930 without a shutdown.

On-line monitoring of cell conditions enables "On Aim" operation.

Analysis of all shipments documents product quality for Du Pont customers.

Both direct current and inductively coupled plasma spectrophotometers allow trace impurity analyses for product and raw materials.

5

Appendix A (continued)

The applications of sodium are summarized in Du Pont Bulletin-Sodium Uses and Advantages.

Sodium is used for the production of organometallic compounds, particularly the antiknock agents tetraethyl and tetramethyl lead.d

The ability of sodium to reduce metal chlorides is utilized in the production of refractory metals (titanium, zirconium, etc.), potassium metal, and silicon. Similar processes have been described for producing magnesium and calcium.

Sodium monoxide and peroxide are made by controlled oxidation of sodium.

Sodium hydride, made by direct reaction of sodium and hydrogen, is used in organic syntheses and to produce other hydrides such as sodium borohydride. Sodium hydride in molten sodium hydroxide is used for metal descaling.

Sodium will react with ammonia at 300 C (572 F) to form sodamide. Sodamide reacts with carbon at 700 C (1292 F) to form sodium cyanide.

Sodium is used in the manufacture of pharmaceuticals, triphenyl methane dyes, synthetic indigo, herbicides, insecticides, synthetic perfumes, detergents, lubricants and waxes.

Sodium catalyzes many polymerizations such as 1,4-butadiene and styrene-butadiene mixtures.

Sodium can be used to promote adhesion of polyfluorinated resins (such as Teflon®) to substrates.

Sodium treatment of petroleum stocks will generally remove sulfur present in forms which are resistant to hydrodesulfurization.

Sodium is used to recover lead from complex lead dross containing lead, copper, antimony, and sulfur. Sodium is utilized to liberate high-value metals from their ores and to purify metals.

Both primary and secondary batteries have been developed in which sodium is an electrode.

Polyethylene-sheathed sodium cables have been used for underground electrical power distribution systems.

A variety of applications of sodium as a heat transfer medium have been developed, most notably in the operation of liquid-metal cooled fast-breeder power reactors and solar energy collector systems.

Sodium requires special storage and handling techniques to avoid fire, explosion, and personal injuries.

HEALTH HAZARDS

Sodium causes severe burns to the skin and eyes by reactive formation of sodium hydroxide. Contact with moist skin is especially hazardous, as severe burns may result from both thermal and chemical effects. Fumes from sodium reactions may irritate the nose, throat and lungs if inhaled. Aside from burns, sodium is not known to cause any systemic or chronic effects.

Sodium metal has a low vapor pressure. Airborne sodium is nom-rally the result of sodium burning and is in the form of sodium monoxide. Du Pont recommends that exposure to airborne sodium be limited to 1 mg/m3, 8-hour time-weighted average. When moisture is present, sodium hydroxide will form. The OSHA limit and TLV® for NaOH is 2 mg/m3. TheTLV® is a ceiling limit, while the OSHA value is an 8-hour timeweighted average.

SAFETY PRECAUTIONS

Persons handling sodium should be aware of its hazards. They should be thoroughly instructed in the proper first aid procedures and should wear appropriate personal safety equipment. They must avoid getting sodium in eyes, on skin, or on clothing.

Sodium reacts with water, forming sodium hydroxide and hydrogen gas. In air, the hydrogen usually ignites, resulting in explosions. Exposed sodium reacts with moisture from the air to form sodium hydroxide and hydrogen gas. If hydrogen gas accumulates in a container or a building, an explosion may occur. The explosive range in air is 4.1 to 74.2 volume percent hydrogen.

Bulk sodium does not explode or detonate. Sodium will autoignite in air above 120-125 C (248-257 F), giving off dense white smoke that reacts with moisture to form sodium hydroxide. Once ignited, sodium burns with a hot flame.

Appendix A (continued)

Exclusion of air by covering or smothering with dry, light soda ash will extinguish a sodium fire.

FIRST AID

Sodium burns are of the same nature as caustic burns and should be treated by immediate and prolonged flushing with cool water. This initial action should be taken before transporting the victim for medical help, to remove the caustic and prevent further tissue damage.

In case of contact with skin, particles of sodium adhering to the body or clothing should be removed immediately. This must be done before washing to avoid additional burns from the heat of reaction between sodium and water. After the sodium has been removed, the affected area should be flushed with cool water for at least 15 minutes. Medical attention should be obtained promptly. Contaminated clothing and shoes should be washed or burned.

If sodium contacts the eyes, they should be immediately flushed with copious amounts of water for at least 15 minutes. A physician, preferably an ophthalmologist, should be called promptly.

If fumes from sodium reactions are inhaled, the victim should be removed to fresh air immediately. If not breathing, artificial respiration, preferably with an oxygen resuscitator (or mouth-to-mouth), should be administered. If breathing is difficult, oxygen should be provided. A physician should be called.

PERSONAL PROTECTIVE EQUIPMENT

Sodium is shipped in solid form only. It is available as drums of molded brick, cast or "fused" drums, and in bulk form in tank trailers, tank cars, and ISO containers.

Solid Sodium Handling - Chemical splash goggles and loose-fitting dry moleskin mittense are recommended for handling sodium bricks. Mittens are preferred to gloves because mittens can be removed more quickly. They should be changed frequently to avoid contamination, and can be washed and dried for re-use. Additional protection may be needed to avoid getting caustic on the skin or clothing. When removing bricks from drums or otherwise handling sodium bricks, rubber aprons and gauntlets for arm protection are recommended. Sodium particles will pick up moisture, forming sodium hydroxide which can cause severe burns to unprotected skin.

Liquid Sodium Handling - Additional protective clothing is required when molten sodium is handled. A face shield should be worn in addition to goggles when protection of the entire face is needed. A face shield should not be used as a replacement for chemical splash goggles. Special chemical splash goggles are recommended: American Optical Special Goggle No. SCS247-711, with cemented-in 0.060 lenses, rubber band, and forehead and nose shields. The nose shield has a hook on the beak which permits effective use of a fire retardant Nomex[d] aramid fiber bandana (diagonally folded 27-inch square piece of cloth) to cover the lower face, ears, and neck. This equipment combination provides eye, face, and neck protection with comfort. Head protection should be provided by a hard hat or flame-proof cap or hood.

Dry moleskin mittens or aluminized Kevlar[d] long gauntlet mittens are used for hand protection so they can be "slipped off" if necessary Gloves are not as safe because they do not permit easy removal. Multi-layer garments of Du Pont Nomex[R] aramid fiber are recommended and have proven far superior to flame retardant cotton or polyester work clothes for preventing sodium burns. Close-weave Nomex[R] with air permeability of less than 100 feet per minute is most effective. Clothing should be designed so that it can be torn or cut off rapidly in case the wearer is splashed with molten sodium. Leather belts should not be worn. Aluminized Nomex[R] aprons or specially designed high temperature suits should be considered. Heavy leather safety shoes and leather and canvas spats, attached with Velcro[R] fasteners, can be worn over the shoe tops if needed.

RESPIRATORY PROTECTION

Sodium at ambient temperature does not form harmful vapors. However, respiratory protection is needed to avoid exposure to hazardous sodium oxide or sodium hydroxide smoke from burning sodium.

Half-face or full-face NIOSH approved respirators equipped with suitable dust filters will afford protection against sodium oxide smoke. These should be cleaned frequently and filters changed when breathing resistance increases. **Filter or dust type respirators do not offer protection against gases, vapors, or an oxygen deficiency. Protection from these requires a source of breathable air or oxygen. Where smoke is very heavy, a full rubber suit with breathable air supply should be worn.** Respiratory protective equipment must be carefully maintained, inspected, cleaned, and sterilized at regular intervals and before use by another person. Consideration should also be given to possible hazards of other substances used in conjunction with sodium.

---

*Moleskin is a special tightly woven cotton material. Consult a Du Pont representative for detailed description of protective equipment.

Appendix A (continued)

Specialized protective equipment avoids injury to Du Pont and customer personnel.

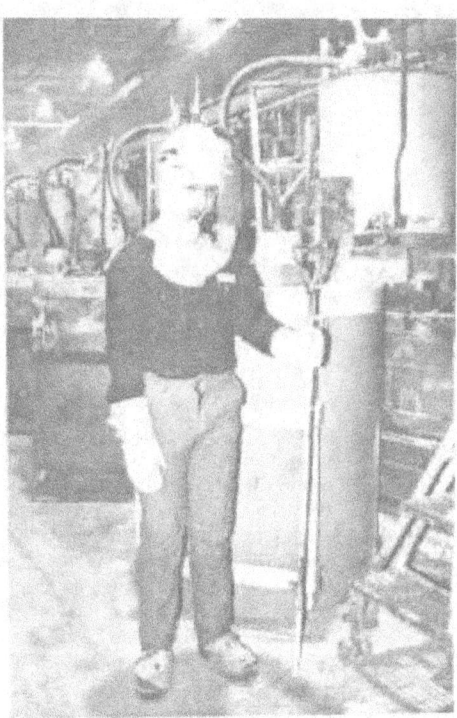

Appendix A (continued)

Sodium fires are extinguishable with dry sodium carbonate (light soda ash) or by exclusion of air. The following materials should not be used on sodium fires, since they react violently with sodium:

- water
- halogenated hydrocarbons, such as carbon tetrachloride
- carbon dioxide

(See "Chemical Properties", p. 3).

Class A, B, and C fire extinguishers should never be used on sodium fires and normally should not even be stored in the sodium area. Class D or dry powder extinguishers containing soda ash are compatible with sodium, provided nitrogen and not carbon dioxide is used to expel the powder. Caution must be used with a pressurized extinguisher because the liquid sodium may be spattered by the discharge from the extinguisher. Salt or dry graphite powder may also be used on sodium fires, but they are less effective than dry soda ash. Signs should be posted at building entrances and throughout the area warning against use of water and standard type fire extinguishers.

Plant and local fire departments which might respond to a fire in the sodium area should be afforded prior and periodic training. These personnel should know where sodium is stored and acceptable techniques to extinguish sodium fires.

Sodium fires burn on the surface of a molten sodium "pool." Fires may be extinguished by smothering, after first containing the liquid sodium to minimize its surface area. Firefighters should stay upwind. Soda ash can be used to dike a sodium spill, and liberal quantities should be spread over the burning surface. If sodium is burning in open equipment such as a drum or tank, sheet steel can be used to smother the fire.

A substantial supply of dry soda ash in properly labeled and covered drums should be stored where sodium is used. A back-up supply of soda ash adequate to handle a major spill should be available near bulk sodium systems. Soda ash can be kept in loosely covered, open head drums to minimize moisture absorption while providing quick access. If periodic inspection shows soda ash is caked, it should be discarded and replaced.

If a shipment of sodium is involved in an accident or emergency anywhere in the Continental United States, it should be reported by a toll-free telephone call to the Chemical Manufacturers Association's Chemical Transportation Emergency Center ("CHEMTREC") in Washington, D.C.

1-(800)424-9300

The information specialist on duty will ask the name and location of the caller, the name of the shipper, the product, the shipping point and destination, what happened, nature of any injuries, weather conditions, proximity to populated areas, etc. He will then give the caller recommendations for controlling the emergency situation until the shipper's specialist can relay help. "CHEMTREC" will immediately advise the shipper of the emergency and one of the shipper's specialists will contact the caller promptly.

If a Du Pont sodium shipment is involved, also call Du Pont collect at:

1-(716)278-5158

A similar emergency response system is in effect in Canada. Any transportation emergency should be reported to CANUTEC, a Canadian government agency by calling:

1-(613)996-6666

Advice on handling emergencies will be provided and the shipper will be notified to provide additional assistance. The Du Pont Canada technical advisor may also be contacted directly by calling:

1-(613)348-3616

All four of these phones are manned 24 hours a day.

Appendix A (continued)

SHIPPING CONTAINERS

Tank Cars-Fused Sodium	80,000 lb (36,000 kg) 100,000 lb (45,000 kg) 125,000 lb (56,700 kg)		
Tank Trucks-Fused Sodium	30,000-36,000 lb (13,600-1 6,300 kg) (Depends on state load limits)		
Steel Drums-Fused (DOT 17E)		470 lb gross (213 kg)	420 lb net (190 kg)
Steel Drums (DOT 17C) (Non-returnable)	12-lb Bricks (5.4 kg)	367 lb gross (166 kg)	300 lb net (136 kg)
	5-lb Bricks (2.3 kg)	387 lb gross (175 kg)	320 lb net (145 kg)
	2½-lb Bricks 1-1 kg)	387 lb gross (175 kg)	320 lb net (145 kg)

DOT HAZARD CLASSIFICATION

While shipping containers are normally padded with nitrogen, inert padding with argon and helium is also available. The DOT classification for sodium is "Flammable Solid". The DOT identification number for sodium metal is UN 1428. Packages must be labeled with DOT "Flammable Solid" and "Dangerous When Wet" labels.

FIGURE 1
Standard Sodium Brick Sizes

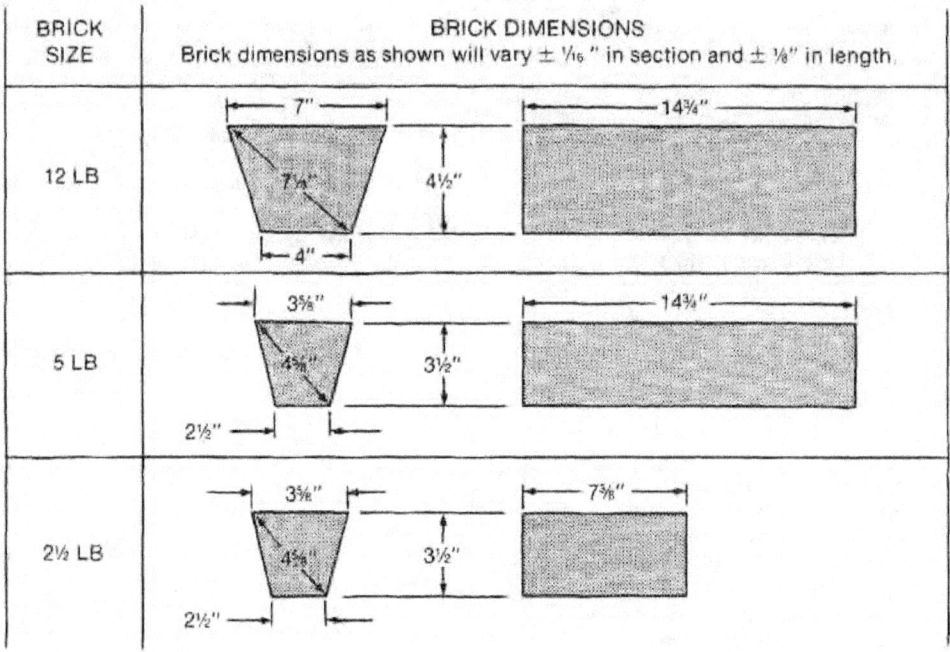

BRICK SIZE	BRICK DIMENSIONS Brick dimensions as shown will vary ± ¹⁄₁₆" in section and ± ⅛" in length.
12 LB	7" / 7¼" / 4" / 4½" / 14¾"
5 LB	3⅝" / 4⅞" / 2½" / 3½" / 14¾"
2½ LB	3⅝" / 4⅞" / 2½" / 3½" / 7⅜"

Appendix A (continued)

PACKAGES

Du Pont sodium is shipped in drums, tank cars, tank trucks, and ISO containers. Steel drums are filled with molten sodium, then cooled to solidify the product. Open head drums are used to ship sodium bricks. Both drums are non-returnable, but arrangements can be made to return empty drums for disposal if the need exists.

Tank trucks, tank cars, and ISO tanks are filled with molten sodium, cooled to solidify the sodium for shipment, and subsequently remelted at the customer site. They are equipped with external channels for oil circulation. Cool oil is circulated after loading and hot oil is circulated for melting.

Each user should consult and follow the current governmental regulations, such as Hazard Classification, Labeling, Food Use Clearances, Work Exposure Limitations, and Waste Disposal Procedures for the up-to-date requirements for sodium.

STORAGE AND HANDLING

Anyone who handles sodium should be thoroughly familiar with the information and recommendations in this bulletin. Du Pont should be contacted for additional information if bulk unloading, storage, and handling of sodium is planned.

MATERIALS OF CONSTRUCTION

Carbon steel is satisfactory for long-term sodium use (300 Series Stainless Steel is also acceptable). Piping should be 2 inch (5.1 cm) diameter, Schedule 40 seamless, with butt welded connections, installed with a minimum number of flanges. Smaller diameter piping is more prone to plugging in use. Gaskets should be compressed white asbestos with Buna S rubber filler as manufactured by the Darcoid Company or equivalent. Teflon gaskets should not be used due to its reaction with sodium.

STORAGE

Sodium should be stored in a dry, fireproof building in sealed metal containers. A remotely located, detached building is preferred. The storage space must be large enough to accommodate both the maximum expected inventory of sodium and any empty containers. The building should not contain flammables, combustibles or water and should not have a sprinkler system, water pipes, steam pipes, or skylights. Particular care must be taken to prevent water entry from roof leaks, rain, snow or improper drainage. Potential for flooding should be evaluated. Small quantities of sodium may not need a separate building or room, but the basic requirements to prevent water contact still apply.

A special bulk storage and transfer system is needed whenever Du Pont sodium is being received in tank cars or trucks. A building is required to accommodate the storage tank and the tank car or truck during unloading. Hot oil facilities for melting rail car and truck shipments are also necessary. The typical hot oil system within sodium tank cars is shown in Figure 2.

A hot oil system capable of circulating 350 gpm (1325 Lpm) oil at 140-145 C (approx. 280-290 F) will melt the sodium in a tank car in 4-5 hours. About 1,500 gallons (5,700 liters) of oil containing appropriate antioxidants for high temperature operation are needed for a bulk system. "Conoco-HT" or its equivalent is essential to prevent oil degradation with subsequent plugging and heating system failure. When melted, sodium is transferred to the nearest receiving point by vacuum or by low pressure, high purity nitrogen. As sodium is withdrawn, the tank car is filled with nitrogen. Procedures for melting the sodium and removing it by pressure or vacuum have been developed by Du Pont. Contact your local Du Pont representative to arrange technical assistance for planning and installing bulk storage systems in accordance with Du Pont's standard practices.

BUILDING DESIGN

Storage tanks and vessels containing sodium should be located indoors in a suitably constructed building and preferably isolated from other facilities or separated by non-combustible walls and floors.

Sufficient exits must exist to permit rapid evacuation of personnel in the event of a spill. No less than two means of egress should be provided from each distinct level, room, or building in which sodium is handled or used. No portion of an area where sodium is handled should be farther than 75 feet (23 meters) from the nearest exit. Additional exits or stairs may be needed based on the number of persons in the building, building height, or other factors. (See the U.S. NFPA Standard #101, "Life Safety Code.")

All exit doors should open out in the direction of travel and should be provided with panic hardware.

Appendix A (continued)

Drums of fused sodium offer convenient supply of intermediate volumes.

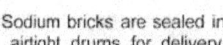

Sodium bricks are sealed in airtight drums for delivery.

Sodium tank cars are loaded in a dry environment.

Tank car dome valving, showing level probes and insulated loading line to right rear angle valve.

12

Appendix A (continued)

FIGURE 2

Sodium Tank Cars are Equipped with Oil Channels for Solidifying
and Melting the Metal.

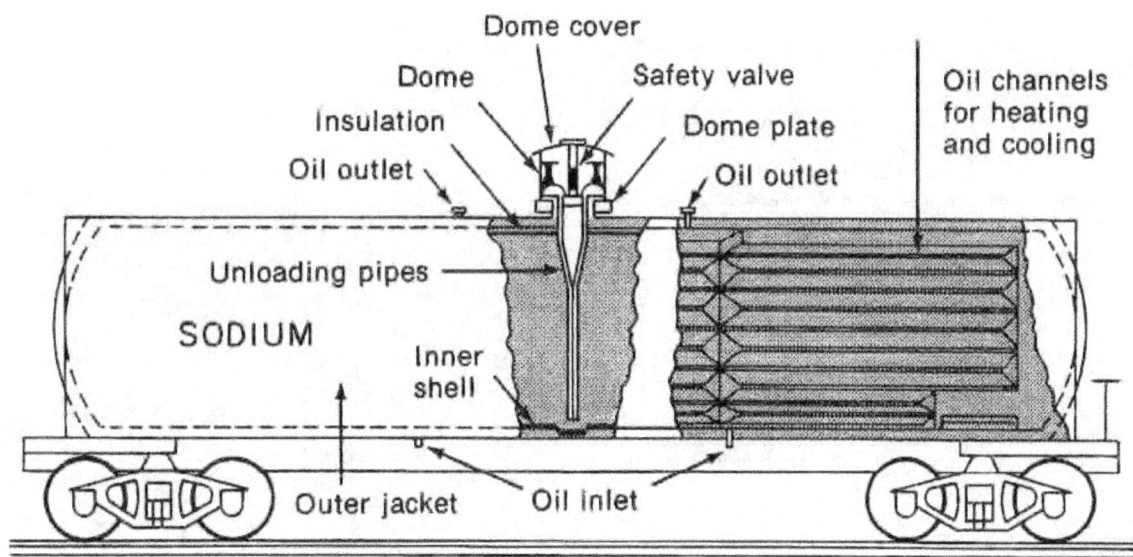

Appendix A (continued)

Natural ventilation should be provided by installing vents, at the highest points to avoid possible hydrogen accumulation.

EQUIPMENT DESIGN

Processes using sodium should be designed and located to avoid all contact between sodium and water. Water piping should be kept away from equipment containing sodium. Floors under such equipment should always be dry and without drains that could accumulate water.

Melting pots and reaction vessels may be heated with hot oil, open flames, electrical resistance, or induction heating, but **never with steam or water.** Equipment for casting or solidifying sodium **must not be cooled with water.**

Sodium may be melted in dry cast iron or steel containers open to the air, if a temperature of about 115 C (239 F) is not exceeded. Depending on air temperature and humidity, molten sodium will autoignite at about 120-125 C (248-257 F). At higher temperatures, sodium must be handled or processed under a refined oil or preferably under an inert atmosphere such as dry, high purity nitrogen. Sodium pipeline temperatures of 113-118°C and storage tank temperatures of 105-110°C are recommended.

In large scale operations, vessels may be charged with liquid sodium. When charging sodium bricks to a closed vessel, a double gate lock (Fig. 5) large enough to hold the sodium bricks may be used. The lock chamber can be purged with an inert gas such as nitrogen, if complete exclusion of air is necessary or if reactor fumes must be purged to avoid escape into the air.

Process equipment should be designed so that explosive hydrogen-air mixtures do not form.

Process equipment should be accessible and permit easy dismantling for cleaning. Inside surfaces of equipment should be made accessible for scraping from the outside through ports or manholes if wall cleaning is needed. Equipment used in sodium process work should have no hollow enclosed sections. Sodium may leak into these spots and present a difficult disposal problem or an unrecognized explosion hazard. Procedures can be developed to permit safe vessel entry for cleaning large tanks such as tank cars and storage tanks, but careful planning is required to do this safely.

To contain spills, a steel pan, curbing, or alternate facilities should be used under sodium equipment. Sodium leaking into the pan is contained and the pan can be covered with a lid or **dry** soda ash in case of fire.

Before placing sodium in any reactor, tank or container, care must be taken to eliminate all water from the equipment. Equipment can be dried by heating every part of the apparatus simultaneously to above 100 C (212 F) with all openings free and then purging with dry air until all traces of moisture have been removed. Alternately, hot dry air can be blown through the equipment until drying is complete.

VENTILATION

Cold sodium does not produce noxious vapors. In areas where cold sodium is handled, general ventilation is sufficient to prevent accidental hydrogen accumulation. In work areas containing sodium oxide smoke, adequate ventilation should be provided to prevent exceeding the sodium hydroxide threshold limit value of 2.0 mg/m³.

AIR ANALYSIS

Routine air analyses for sodium are not usually needed where solid sodium is used unless potential exists for smoke from burning sodium. Equipment and employee training to ascertain any hydrogen presence during emergency spill situations should be considered.

ELECTRICAL EQUIPMENT

Sodium may be handled in "General Purpose" classified areas as defined in the "National Electric Code".

All personnel should be familiar with Safety Regulations and follow them carefully. Operators must be provided with full protective clothing. (See p. 7.) Special attention should be focused on avoiding any chance water contact (leaks, sprays, sprinklers) during any sodium transfer.

Drums of sodium are stored unopened until used. The proper conditions for storing drums of sodium bricks and fused sodium are given under "Storage," page 11.

It is essential that all tools used with sodium and all surfaces which will be in contact with sodium be cleaned and dried. This is discussed under "Equipment Design," page 14.

HANDLING MOLTEN SODIUM

Molten sodium may be transferred by vacuum, nitrogen (or other inert gas) pressure, pumping, or by gravity. There are advantages and disadvantages to each method. Vacuum transfers reduce spill potential. Also, rodding of lines is less hazardous than in pressure systems since blowback from pressure pockets is less likely. However, with vacuum, air entry into the system is more common, which increases solids (sodium oxide) formation and line plugging. When pressure transfer is used, the gas pressure should be maintained as low as practical; 10-15 psig is usually satisfactory. Where gas pressure transfer is used, it is essential to provide safeguards to prevent over-pressurization and for rapid

Appendix A (continued)

FIGURE 3

Drums of Fused Sodium Are Placed in Special Heating Equipment to Liquefy the Sodium for Unloading into the Plant's Receiving System.

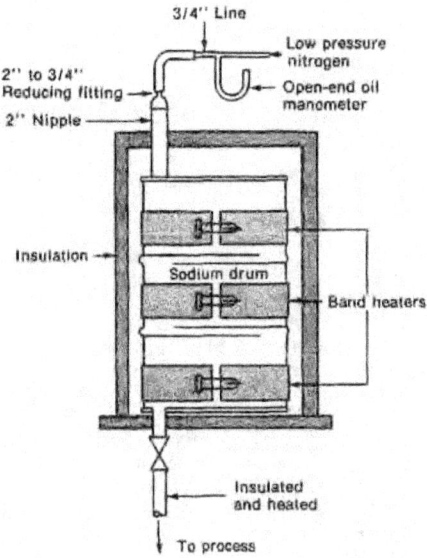

depressurization. Pumps are useful if high pressures are needed for reactor feed. Gravity is the simplest transfer method, but has limited application. In all cases, transfer at temperatures below the auto-ignition temperature (120 C) is very important. A sodium spill without a fire can be cleaned up easily, but burning sodium is destructive and difficult to control.

DRUM MELTING ARRANGEMENT

With electrically heated drum melters, it is possible to melt the contents of a 55-gal. fused drum in five hours with a power input of about 2KW. A partial list of manufacturers follows:

OHMTEMP
P.O. BOX 85-A
GARDEN CITY; MI 48135
TEL. (313) 261-7036

TRENT INC.
201 LEVERING AVE.
PHILADELPHIA, PA 19127
TEL. (215) 482-5000

PALMER INDUSTRIAL HEATING EQUIPMENT, INC.
P.O. BOX 265-T
PLYMOUTH, MI 48170
TEL. (313) 453-8300

The 55-gal. (420-lb.) DOT 17-E nonreturnable drums have both a 2" and a 314" bung on the top and a 2" bung on the bottom in line with the 2" top bung. (Fig. 3). All three of these bungs have standard NPT threads. In preparation for melting, a short pipe nipple, threaded on one end and fitted with a flange on the other is threaded into the bottom 2" bung opening. A thermowell may be threaded into the 3/4" bung opening. A nitrogen line is connected to the top 2" bung opening. Nitrogen is used for displacement to avoid exposing the sodium to air. **This nitrogen line valve should be open prior to and during heating with no more than an ounce per square inch pressure (open-end oil manometer) as a safety measure because drums are not built to be pressurized. After set-up and check-out, the drum can be heated (Figure 3).**

The flange on the pipe nipple is bolted to the flanged valve on the permanent piping leading to the receiving tank. A plug-cock type valve is recommended. The valve and permanent piping should be insulated and heated with strip or wire resistance elements. The short connector can be insulated or, if necessary, can be heated. For indoor piping runs with good (i.e., min 4") insulation, elements producing 25-30 watts per foot of pipe and 50-75 watts around valves are adequate.a It is important to provide temperature control for operation below the sodium auto-ignition temperature of 120 C. When the sodium drum is empty, the connector is stripped of temporary insulation and washed and dried or stored under kerosene until the next use.

CAUTION: **It is essential that the connector be moisture-free before reuse. Explosions and pipe rupture have occurred when proper cleaning and drying procedures were not used.**

aDifferent heat tracing specifications may be required for bulk liquid sodium piping systems due to length or outdoor conditions

Appendix A (continued)

There is a 3-inch freeboard in the top of the drum. When the thermowell is inserted beyond this 3-inch gap, a reading of 110-115 C (230-239 F) on the thermometer indicates that the sodium has melted and is ready for transfer to the receiver. Gravity flow is recommended since the drum is too light for pressurizing. The receiving tank should preferably be on a scale, so that both the emptying of the drum and transfer to the system can be monitored. This tank should also be heated, insulated, and equipped with a temperature recording/alarm instrument and other appropriate instrumentation. If the tank is a properly rated pressure vessel, sodium may be transferred from this tank to the system either by inert gas pressure or by pumping. Piping runs should be as short as possible and joints welded. Rodding of plugged lines can be a hazardous operation, and care should be taken to relieve pressure on the line before rodding. Appropriate shields must be used to deflect any sudden discharge of liquid sodium from the line being rodded. See page 20 for empty drum and waste disposal.

HANDLING SOLID SODIUM

When removing sodium bricks from 280-310 lb steel drums, the nut which holds the lock rim in position is unscrewed, the lock rim removed, and the cover pried off. Clean, dry, non-sparking tools should be used. The drums are purged with inert dry nitrogen when packed. However, if moisture is present during packing or from any subsequent opening, sodium will liberate hydrogen which may form an explosive mixture with air in the confined space. For this reason, it must always be assumed that hydrogen may be present when a container is first opened. Hammering or any other operation which might cause a spark and ignite the hydrogen must be avoided. When opening a container the first time, the cover should be lifted cautiously. When opening or closing a sodium container, the operator should always wear protective clothing, including goggles and full face shield and dry hand protection, preferably moleskin mittens.

In transferring sodium from the shipping container to a distant point of use, a special metal container, equipped with a tight-fitting cover, should be used. Immersion in mineral oil may help in safe transfer/storage. Since sodium picks up moisture when exposed to humid air, the containers must be kept closed, and preferably purged with dry nitrogen except when sodium is being removed.

For safety and convenience, iron tongs may be constructed for handling sodium bricks. The tongs shown in Figure 4 are designed for handling 5-lb. bricks.

Various charging gates have been designed for feeding solid sodium into reactors. A double gate with an intermediate chamber large enough to hold the sodium charge is used when it is important that no air be admitted to the reactor or where burning might occur when the

reactor is opened. If necessary, the intermediate chamber of the gate can be filled with an inert gas (See Figure 5.)

Disposal of waste sodium and empty drums is described on page 19.

SPECIAL FORMS OF SOLID SODIUM

The reactivity of sodium increases markedly with subdivision. In many operations it is necessary to charge portions of sodium smaller than the standard bricks. Where cutting is necessary, it should be done with a steel knife on a sheet iron surface. If cutting is a regular operation, the sodium knife shown in Figure 4 is recommended.

Castings

Sodium bricks can be easily melted and cast into special sizes and shapes in molds made of steel or cast iron. Clean liquid sodium will not burn in air unless the temperature exceeds 120 C (248 F). It is very fluid at 110 C (230 F) and can be easily poured. The heating, melting, and molding equipment should be mounted in a shallow steel pan large enough to hold all of the equipment and sodium being handled at any time. The pan must be kept dry. Spilled sodium should be covered with dry soda ash and removed to a waste container.

The melting pot may be made of steel, cast iron, or from a piece of standard pipe with a plate welded on one end to form a closure. The pot may be provided with metal handles and a pouring spout. The top of the pot should be flat so a metal lid can be laid over it. In closed, dry containers, sodium quickly forms its own inert (nitrogen) atmosphere.

Molds may be of steel or cast iron fabricated to the desired shape. Split molds may be used. The faces should be at least an inch wide and machined so that they form a tight joint when clamped together. Each half of the mold should have a base wide enough to hold it upright when set on the casting table.

The equipment and tools should be heated in an open flame to at least 120 C (248 F) before casting is started; between operations, they should be thoroughly cleaned. (See "Tank and Equipment Cleaning," p 18.)

The melting pot may be heated by open flame, hot oil, electrical resistance, or induction, but never by steam or water. The brick is placed in the pot and a flat iron cover placed on the pot to exclude air. Since sodium is an excellent heat conductor, the brick heats evenly and melts suddenly when the melting temperature is reached. Readings should be taken frequently until the temperature reaches 105-110 C (221-230 F).

A stainless steel tubular dial thermometer should be used. If the castings are to be substantially free of oxide,

Appendix A (continued)

the surface of the molten sodium should be skimmed with a metal spatula or flat steel blade.

The mold is dried by heating to at least 120 C (248 F) and then placed upright in the safety pan and allowed to cool below 90 C (194 F). It may be desirable to secure tall molds to some rigid fixture to avoid the danger of tipping. Molten sodium is poured into the mold and the surface skimmed with a metal spatula or blade. Sodium solidifies rapidly. When the top of the casting is solid, the casting is transferred immediately to an air-tight container.

Any sodium remaining on the mold should be scraped off, and the mold wiped with a small amount of kerosene before it is reused. All sodium waste should be kept in a covered container and removed frequently to the burning area for disposal.

Extrusions

Rods, ribbons, or wires of sodium can be prepared by extrusion in proper equipment. In extruding sodium, precautions must be taken to avoid oxidation of the surface.

Shot and Pellets

For some chemical reactions there are advantages in using metallic sodium as "shot" or "sand" rather than in massive form.

One method of preparing sodium "shot" is to melt bricks under dry hydrocarbon oil, toluene, xylene, or any high-boiling organic liquid inert to sodium and previously refined with a small amount of sodium. The liquids are then agitated. Controlling oil temperature and rate of agitation yields small droplets of molten sodium which freeze as discrete particles when cooled.

Where the presence of the dispersing liquid used to prepare the finely divided sodium does not interfere with subsequent operations, this suspension may be added directly to the reaction vessel. If sodium in this form is required free from dispersing liquid the solid particles can be washed with a lower-boiling solvent and this solvent removed by evaporation in a nitrogen atmosphere. When handling finely divided sodium, it must be remembered that it is extremely reactive and caution is needed to prevent losses through air oxidation.

Finely divided sodium can be dispersed in molten paraffin and the solidified blocks cut and charged into the reactor. In other cases the suspension of sodium in liquid paraffin can be used.

Sodium can also be cut into small pieces. The pieces should be placed in an air-tight container or otherwise protected against surface oxidation.

Dispersions

Methods have been developed for producing dispersions of sodium in small particle sizes. The greatly extended surface area of sodium in microscopic particles makes possible reactions which are impossible or impracticable with large pieces.

A suspension of sodium in an inert medium which has a boiling point above the melting point of sodium is passed through a high-speed agitator having strong shearing action. Very small amounts of addition agent which help to control particle size, viscosity, and stability are used. Sodium dispersions prepared with up to 50% (by weight) of metallic sodium are free-flowing and permit feeding and metering by gravity, pressure, or pumping.

FIGURE 4a

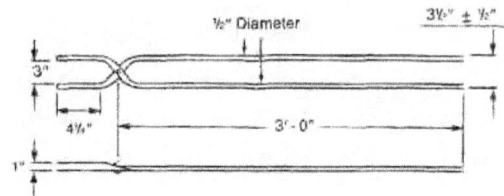

Figure 4a - Iron tongs can be used for handling sodium bricks.

FIGURE 4b

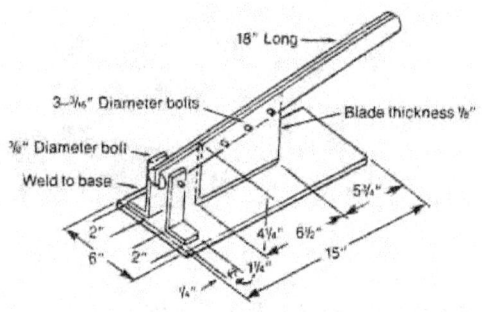

Figure 4b - Sodium cutting knife is all-steel construction. For heavy service, thickness of blade is increased to 1/4".

Appendix A (continued)

High-Surface Sodium

An extremely thin layer of sodium on an inert, free-flowing solid such as soda ash, colloidal carbon, anhydrous activated alumina, or sodium chloride, provides metallic sodium in an extremely reactive form. High yields and reaction control in certain applications are obtained with high-surface sodium.

FIGURE 5

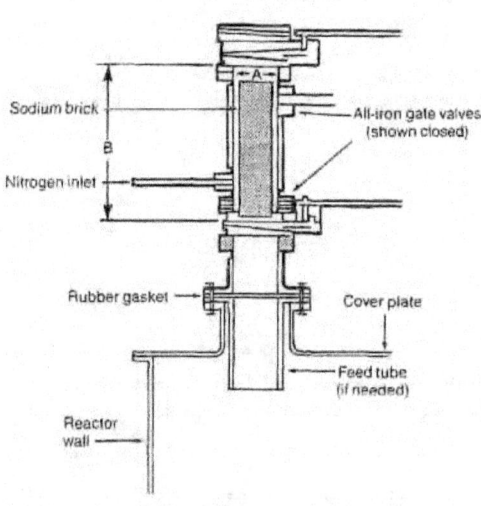

Figure 5 - Double charging gate for feeding solid sodium into reactors.

Brick Size	A	B
12 lb	8	16
5 lb	6	16
2'12 lb	6	9

TANK AND EQUIPMENT CLEANING AND REPAIRS

Process equipment cleaning should be done by persons fully familiar with the hazards and the safeguards necessary for safe performance of the work. Consult Du Pont if cleaning of sodium storage tanks is required.

Proper personal protective equipment such as goggles and face shields, protective clothing, and respiratory protection should be worn when inspecting or cleaning sodium process equipment.

CAUTION: **Precise plans should be made before handling or temporarily storing sodium scrap and equipment containing residual sodium. Sodium should not be "contained" in any manner which may allow the accumulation of evolved hydrogen. Upon reaching the lower explosive limit, the hydrogen may explode.**

Sodium equipment to be cleaned is first drained of liquid sodium. It is then cooled, and accessible solid residues are removed. Finally the equipment is dismantled and as much as possible of the remaining solid sodium is removed. Some sodium residues are highly pyrophoric and air contact may ignite them.

Wherever possible, equipment to be cleaned should be taken to a cleaning area away from the operating area. Small equipment to be cleaned should be placed in a dry steel pan for transporting. Equipment must not be placed on gravel, cinders, or other surfaces where water may be encountered. If the equipment is large, it should be designed so that every part can be purged with inert gas to remove hydrogen.

If the residues contain materials other than sodium or its air oxidation products, a small sample should be tested for abnormal reactivity before cleaning is begun. If any doubt exists as to the safety of the proposed cleaning operation, consult the Du Pont Technical Representative.

Sodium residues are removed by scraping or chopping from cooled equipment after purging with dry nitrogen. Air hammers may have to be used in certain cases. The hammer chisel should be of spark resistant material if volatile flammables are present. Equipment should be completely dismantled so that every part of the interior surface can be scraped. Scraping from the outside through ports or manholes is desirable. If the equipment is too large to make this practical, tank entry may be required. This is a hazardous operation and careful planning is required.

After the bulk of the sodium has been removed, the remaining residue may be removed from the equipment by burning. This should be done in an area where the dense white sodium monoxide smoke evolved can be handled safely. Care must be taken to avoid contact with these fumes. (See "Waste Disposal.") Where appropriate, 150 C (302 F), dry air can be blown against the sodium residue to oxidize it. Alternately, an open flame may be directed against the residual sodium.

In many cases it is difficult to completely burn sodium. Therefore, to destroy all sodium and dissolve the oxide, the equipment is finally washed with water (or steamed). This can be accomplished by placing the equipment in an open dry tank in an isolated and well-ventilated area in such a manner that no evolved hydrogen can be trapped. The tank is then filled with water which is agitated with steam. Steam and water valves should be located a safe distance from the area, which should not be entered until the equipment has been submerged for one hour.

Appendix A (continued)

If final washing with steam or water from hoses is planned, extended lances, protective shields, or other alternatives should be provided for worker protection.

Equipment used with sodium should never be relegated to scrap areas without cleaning.

Welding or cutting on equipment containing sodium can be done, but extensive preparations are required. Each job must be planned in detail before it is started. One technique successfully used for pipe repair or alterations is to strip insulation and heaters to permit freezing of the sodium with air blown on the outside of the pipe. The cooled pipe (and sodium) is then removed with cold cutting tools. Next, the sodium in the line is dug out in both directions as far back from the cuts as possible, and a new section of pipe is welded in place. Air should be blown on the outside of the pipe during the welding operation to reduce pipe temperature and avoid melting the sodium. These jobs should never be attempted with pressure on the system.

Use of nitrogen with the above technique may make equipment repair safer in some areas, but high temperature methods should not be used where flammable organics or hydrogen are present, or danger of explosion exists. Also, care must be taken to make sure nitrogen use does not reduce oxygen content of the air to an unsafe level for breathing.

WASTE DISPOSAL

Sodium waste can be destroyed by burning, reacting with water or steam, or by chemical reaction with alcohols. Special planning and care must be used in disposal to avoid personal exposure or environmental problems.

Small quantities of sodium can also be allowed to "weather" in the open where the sodium can react harmlessly with moisture to form caustic and release hydrogen. Weathering should be monitored to avoid danger to workers and provisions must be made for eventual caustic disposal. Sodium weathers slowly, particularly in dry air, and a coating of oxide, caustic, or carbonate may effectively protect sodium in the interior of large lumps.

Disposal of sodium using a combination of steam, nitrogen and water can be safely handled in an appropriate location. Dry steam or dry steam plus nitrogen will dilute air and hydrogen off-gas and prevent or moderate hydrogen explosions. Again, provisions for caustic disposal are required.

For burning, provisions must be made for emissions of sodium monoxide fumes. The Hauck Safety Torch (Hauck Mfg. Co., P.O. Box 90, Lebanon, PA 17042) may be applicable. Disposal of a large quantity of sodium by burning is usually feasible only in a burning chamber equipped with an appropriate scrubber. Relevant regulations should be reviewed before sodium disposal.

Small quantities of sodium residues such as those encountered in the laboratory can be conveniently destroyed by reacting the material with excess alcohol. Ethanol or isopropanol are preferred over methanol because they are less volatile and give a slower reaction rate. Since hydrogen is evolved in the reaction between sodium and alcohols, care must be taken to avoid an explosive accumulation of air-hydrogen-alcohol mixtures. The vessel should be purged with nitrogen before and after the reaction, with provisions for safe venting of hydrogen formed during the reaction.

DRUMS AND EMPTY DRUM DISPOSAL

Once empty, <u>fused</u> sodium drums (DOT 17E) should be promptly removed from the drum heater. A drum cutter is used to cut off the heads and split open the cooled drums. This allows the drum to be flattened, inside up. Residual sodium is scraped off, for use or waste disposal as described above. The entire drum surface can be coated with kerosene or machine oil before transport to the burning area. Before disposal of the flattened drum, burned metal surfaces should be thoroughly washed and dried.

Once empty, sodium brick drums (DOT 17C) should be cleared of sodium residues, then steamed by means of an open steam nozzle set in a floor grating over which the empty drum can be inverted. The steamed drum should be flushed with water and dried before disposal.

Appendix A (continued)

E. 1. du Pont de Nemours & Co. (Inc.)
Wilmington, Delaware 19898

U.S. Sales and Services

For placing orders or requesting additional product information,
please use our convenient 24-hour toll-free telephone number.
If you prefer, you can write to us.

By Phone

Toll free in continental U.S. (except Delaware)
(800) 441-9442

In Delaware
(302) 774-2099

By Mail

E. I. du Pont de Nemours & Co. (Inc.)
Chemicals and Pigments Dept.
Customer Service Center
Wilmington, DE 19898

International Sales Offices

CANADA
Du Pont Canada Inc.
Box 660
Montreal S, P.Q. H3C 2V1
(514) 861-3861

Du Pont Canada Inc.
P.O. Box 2300
Streetsville Postal Station
Mississauga, Ontario L5M 2J4
(416) 821-5570

LATIN AMERICA
E. I. du Pont de Nemours & Co. (Inc.)
Chemicals and Pigments Dept.
Latin America Sales Office
Brandywine Building
Wilmington, DE 19898
(302) 774-3403

EUROPE
Du Pont de Nemours International S.A.
P.O. Box
CH-1211
Geneva 24, Switzerland
022-378111

ASIA-PACIFIC
Du Pont Far East, Inc.
Kowa Building No. 2
11-39 Akasaka 1-chome
Minato-ku
Tokyo 107, Japan
5855511

Du Pont Far East, inc.
Maxwell Road
P.O. Box 3140
Singapore 9051
273-2244

Appendix A (continued)

NRC INC.

STANDARD OPERATING PROCEDURE

PLANT: 2	AREA: REDUCTION AREA	NO.:2116-15-0221

SUBJECT: Sodium Barrels	REVISION:
	DATE:09/10/91
	SUPERSEDES:
	PAGE: 1 OF: 5

REVISIONS:	APPROVAL
	QA MANAGER:
	PLANT SUPT.:
	AUTHOR (RESP.):

I.   SELECTING AND MOVING SODIUM BARRELS

    A)   There are two types of sodium:

        1)   Black drums/gray drums - for J, M, V, and VM runs (Technical Grade).

        2)   Blue drums/red drums - for V3S, Z, H1, H2M, KFXX and YM runs (Reactor Grade).

    B)   Once the proper drum has been chosen, carefully and thoroughly inspect the drum for any physical damage. Report any damage to the supervisor. Remove any seals and loosen bungs with bung wrench. Use a chainfall to bring a new sodium barrel to the drum warmer. Make sure drum is securely clamped before transporting. Always keep drum close to ground level as it is being transported:

II.   INSTALLING SODIUM BARREL

    A)   Notice at the bottom of the drum warmer there is a groove or half ring. It's purpose is to keep the sodium barrel from sliding due to the incline of the platform.

FORM 2136-009

Appendix A (continued)

B) When lowering sodium barrel to drum warmer:

    1) Place bungs to their proper position for Na/argon inlet assembly hookup. 2" IPS bung for Na bayonet - 3/4" IPS bung for argon inlet.

    2) Lower sodium barrel over the half ring. Make sure drum is in place and cannot slide downward.

C) Close drum warmer and latch.

D) Hook up argon

    1) Remove small bung from top of drum.

    2) Screw in clean argon inlet/TC well assembly (TC should be installed).

    3) Hook up argon supply line to argon inlet assembly.

E) Hook up sodium

    1) Remove large bung from top of drum.

    2) Screw in clean dry bayonet.

F) Flush with argon

    1) Remove union from top of sodium bayonet.

    2) Fully open argon valve.

    3) Check open end of bayonet for flow. If no flow, turn off argon and check lines for plugs.

    4) Once flow is established, flush for 2 minutes.

    5) TURN OFF ARGON. This must be done, otherwise sodium will surge from drum when drum is warm and union is removed to make final connection to accumulator fill line.

    6) Install new "O"-ring on sodium bayonet and replace union.

G) Insulation: Install sodium drum cover (cover all exposed areas).

III. CONNECTING SODIUM DRUMSIDE LINE

A) Turn sodium barrel heater on per instruction in Accumulator Procedure So. 2116-15-0222.

Appendix A (continued)

B) As the sodium (Na) melts, both the bayonet and TC well lowers into the drum.

C) When sodium in barrel is completely melted, push bayonet and TC well to the bottom of the barrel. Gently tap the top of the union of the bayonet to make sure you are at the bottom of the drum. Make sure packing glands are tight.

D) Before removing union from sodium line, check that argon valve to sodium drum is closed.

E) Remove union from sodium line and make connections to accumulator fill line.

F) Wrap insulation around sodium line.

IV. SODIUM BARREL DISCONNECTION

A) Shut off power to barrel heater and to electric tracing of lines by barrel. Open drum warmer to expose sodium barrel. Remove insulation and let cool.

B) Check that barrel is cool to touch before proceeding. If still warm, let drum cool before handling.

C) Shut off argon flow valve.

D) Disconnect argon line and remove TC.

E) Disconnect sodium line: cap both ends to keep moisture and oxygen from entering. Do not remove the sodium bayonet line from the drum. Set empty drum with bayonet aside for future sodium waste disposal (See Sodium Drum Cleaning below).

F) Pull barrel out from drum warmer and set on floor.

G) Unscrew argon inlet and TC we'll assembly. Install bung in barrel. Wash argon inlet with water in hood room. When washing argon inlet and TC well assembly, observe the safety rules for cleaning sodium with water.

H) After washing, dry the argon inlet and TC well assembly with acetone. Pour acetone on both inside and outside of the assembly.

Appendix A (continued)

GENERAL SAFETY INFORMATION

Sodium (Na) and NaK react violently with water. If air is also present, the heat of this reaction can ignite the hydrogen-oxygen mixture and cause an explosion. The severity of the explosion is related to the quantity of Na and NaK and water in contact and can vary from "popping" to a violent destructive blast. The purpose of these instructions is to insure that unintentional contact with water is not made while handling these materials and that when intentional contact occurs, personnel are not in a position to be injured by the reaction.

V.   SODIUM DRUM CLEANING:

Protective Equipment - Face shield, Safety Glasses, Gloves

A)   Shut off all water sprays and hoses into washroom.

B)   Put DRY tray on grate.

C)   Set barrel cradle in tray - leaning back towards wall.

D)   Dehead the empty sodium drum.

E)   Place head against back wall to wash when drum is washed.

F)   Place deheaded sodium drum in cradle with bungs and seams away from low point as cradled.

G)   Start scrubbers.

H)   Insert torch nozzle through door and ignite.

I)   Close both doors securely - wear dark burning glasses.

J)   With flame at moderate to low setting, melt sodium into a pool at the low point of the tilted drum. Begin melting sides and work towards rear of drum. If torch keeps going out, use a smaller flame.

K)   When all sodium has formed a pool, apply excessive heat to pool.

L)   Turn off flame and withdraw through door. Sodium will burn by itself until fully oxidized. Stirring the partially oxidized pool will expedite the process. ALWAYS BE SURE THAT THE STIRRING TOOL IS DRY, OTHERWISE A VIOLENT REACTION WILL RESULT.

Appendix A (continued)

    M)   After burning is complete, let drum cool then lift drum (with heat resistant gloves) and place in corner of washroom.

    N)   Secure doors and slowly wash drum and head.

VI.   REMOVING AND CLEANING SODIUM BARREL BAYONETS:

Protective Equipment - Face Shield, Safety Glasses and Gloves

    A)   Remove bayonet from empty cold sodium barrel using pipe chain wrench. Install bung on barrel.

    B)   Carry bayonet to hood room. Make sure all water sprays and hoses are off.

    C)   Clamp bayonet in vise. Remove bayonet cap and poke s/s rod into bayonet.

    D)   Secure hood room doors and turn on scrubbers.

    E)   Slowly wash bayonet with water hose. Try to keep sodium burning. Use small amounts of water to keep reactionn going.

    F)   When bayonet is clean, flush off inside and outside with water. Then pour acetone through it to dry it off.

    G)   Place clean bayonet in rack provided in sodium barrel area.

APPENDIX B

Problems Relating to Fire Service Awareness of SCBA Fire Resistance Standards

The current standard for self-contained breathing apparatus (SCBA) for firefighters is NFPA 1981 (1992 edition). Prior to 1987 there were no fire resistance requirements for SCBA units in the NFPA standard. Limited fire resistance requirements were introduced in the 1987 edition. The 1992 edition of NFPA 1981 introduced a function test with full flame exposure. All SCBA units must also be approved by the National Institute for Occupational Safety and Health (NIOSH); however, NIOSH does not test SCBAs for fire exposure.

The Scott model 4.5 and model 2A SCBA units that were in use in Newton were produced before the current edition of NFPA 1981 was adopted. The model 4.5 units that were in use in Newton are believed to have been manufactured between 1987 and 1991, while the model 2A units were manufactured prior to 1987. The facepiece of the model 4.5 units was secured to the user's head with nylon composite straps and a polyester knitted cap, and the body harness straps on the model 2A units were made of polypropylene material. These materials were not required to pass fire exposure tests at the time the units were manufactured; however, new units are manufactured with materials that meet the current flame exposure test standards.

Scott's designs of both models were upgraded in 1987 to comply with new requirements of NFPA 1981. The designs were further upgraded to comply with the 1992 edition. The upgrade from 1987 to 1992 involved several changes. To bring the model 4.5 SCBA into compliance with the 1992 edition of NFPA 1981, the model AV-2000 facepiece was introduced. The new facepiece has several upgraded components, including a Kevlar® neck strap and head harness assembly. Modifications were also made to add speaking diaphragms and built-in nosecups and to reinforce the facepiece frame.

The body harnesses on all Scott SCBA models were upgraded to a fire resistant material to meet the 1987 standard and a more fire resistant grade of Kevlar® was introduced in 1992. The regulator airflow capacity was also increased on the newer units.

Upgrade kits have been available from Scott to bring pre-1987 units up to that standard, and similar kits are being developed to allow for existing units to be upgraded to meet the 1992 edition of NFPA 1981.

The non-fire resistant materials are still available for customers who do not have to comply with NFPA 1981, such as industrial users who do not engage in firefighting.

The literature accompanying the new AV-2000 facepieces states that two material options are available for the facepiece harness assemblies.

Model P/N 804177-01 with the flame resistant Kevlar® materials
Model P/N 804063-01 with the polyester and nylon composite materials (These are the materials that were available prior to 1992.)

Appendix B (continued)

The following warning is printed in bold type immediately below this information:

WARNING

DO NOT USE HEAD HARNESS SCOTT P/N 804063-01 FOR FIRE FIGHTING OR WHEN EXPOSURE TO HIGH HEAT OR FLAME IS POSSIBLE. THIS HEAD HARNESS INCLUDES MATERIALS WHICH MAY MELT OR BURN WHEN EXPOSED TO FLAME OR RADIANT HEAT.

A Scott Aviation representative confirmed that the company has not issued any notice to fire departments to inform them of the information contained in this warning, other than to include it in the literature provided for new units. The company's stated position on this issue was that there is no requirement in the NFPA system to upgrade units produced prior to the effective date of new editions of a standard; they are only required to comply with the edition that was current at the time of production. Older units are not required by NFPA to be upgraded to meet the current edition of NFPA 1981.

The representative stated that it is up to each fire department, regulatory agency, or other authority having jurisdiction to determine the particular edition of the NFPA standard that must be complied with and whether or not any particular units need to be upgraded.

The Scott representative further stated that it is the responsibility of each fire department and regulatory authority to stay current with the requirements of the applicable NFPA standards and other regulations, and that it is up to fire departments to determine if their units should be upgraded to meet new standards. The representative noted that many calls are received from fire departments asking about the need to upgrade different units to comply with standards and that many of the callers appear to be unaware of the changes in regulations, which standards they are required to comply with, and what changes would need to be made to bring units into compliance with newer standards. (Information provided by Linda Strawn, Customer Service Representative, 800-247-7257.)

APPENDIX C

Photographs

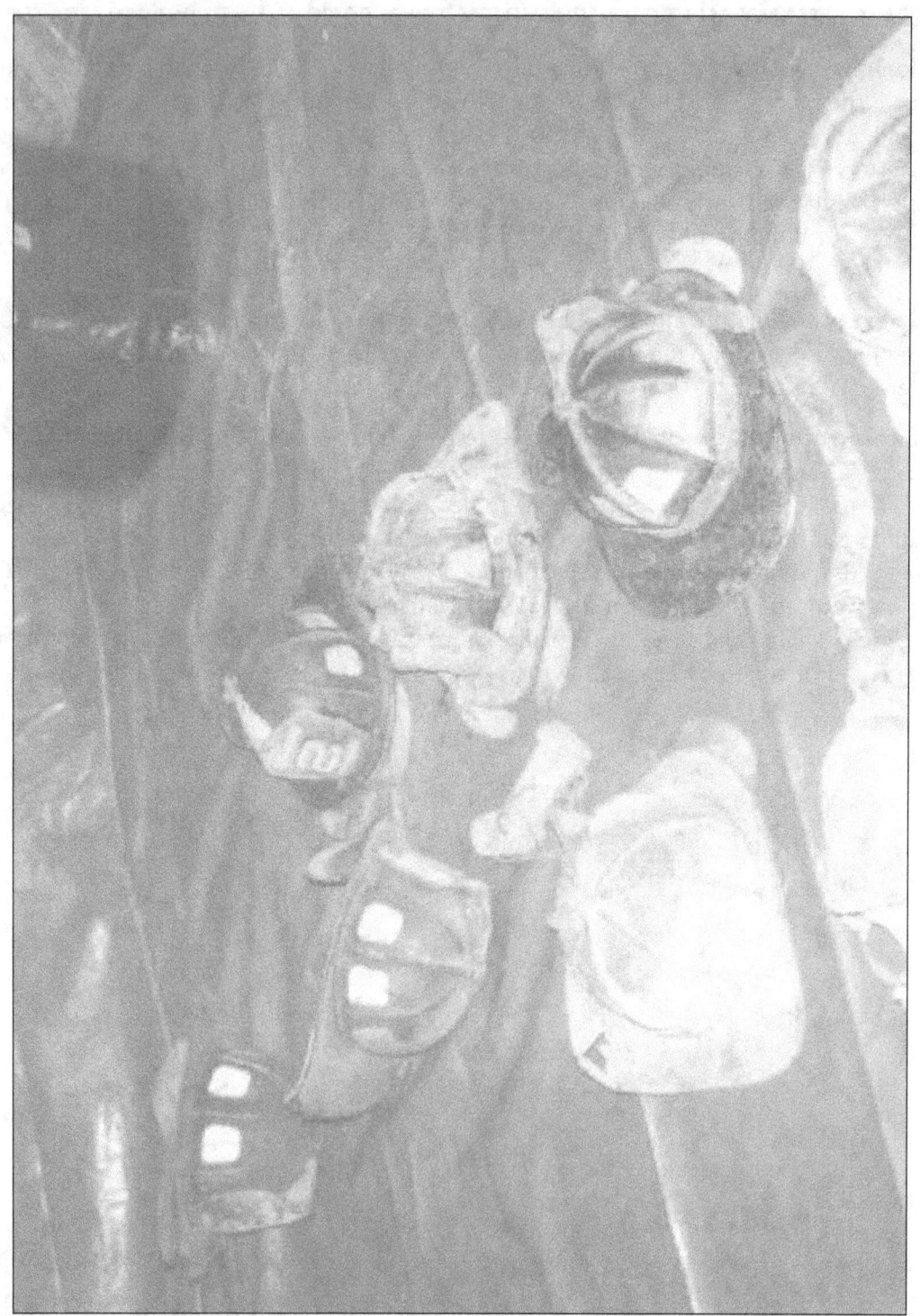

Photo by J. Gordon Routley

Helmets worn by injured personnel show different degrees of damage. Most were worn without chin straps, which allowed the force of the explosion to lift the helmets from the users' heads.

Appendix C (continued)

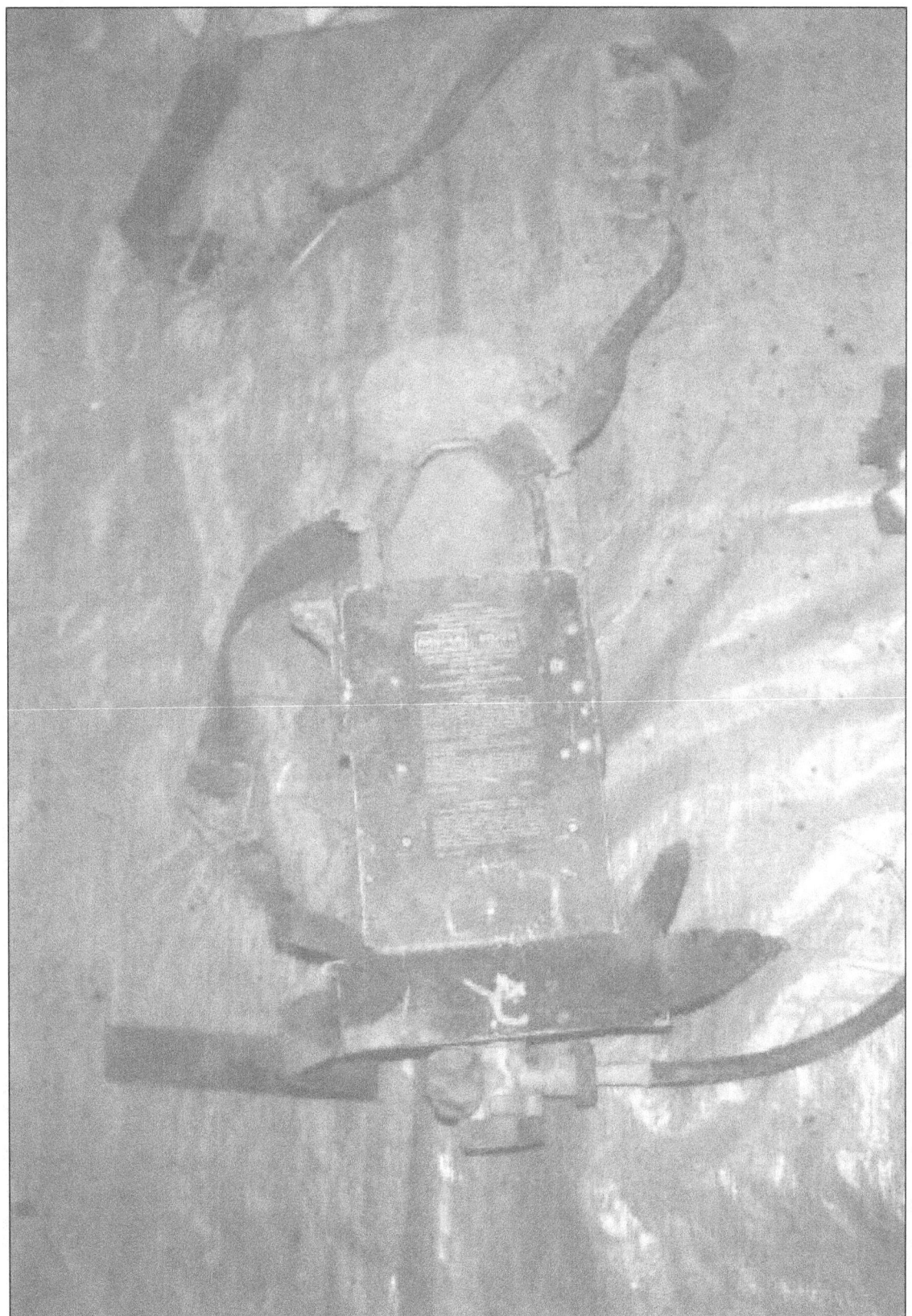

Photo by J. Gordon Routley

Backpack from Scott 2A breathing apparatus showing burn through of waist and shoulder straps on the left side.

Appendix C (continued)

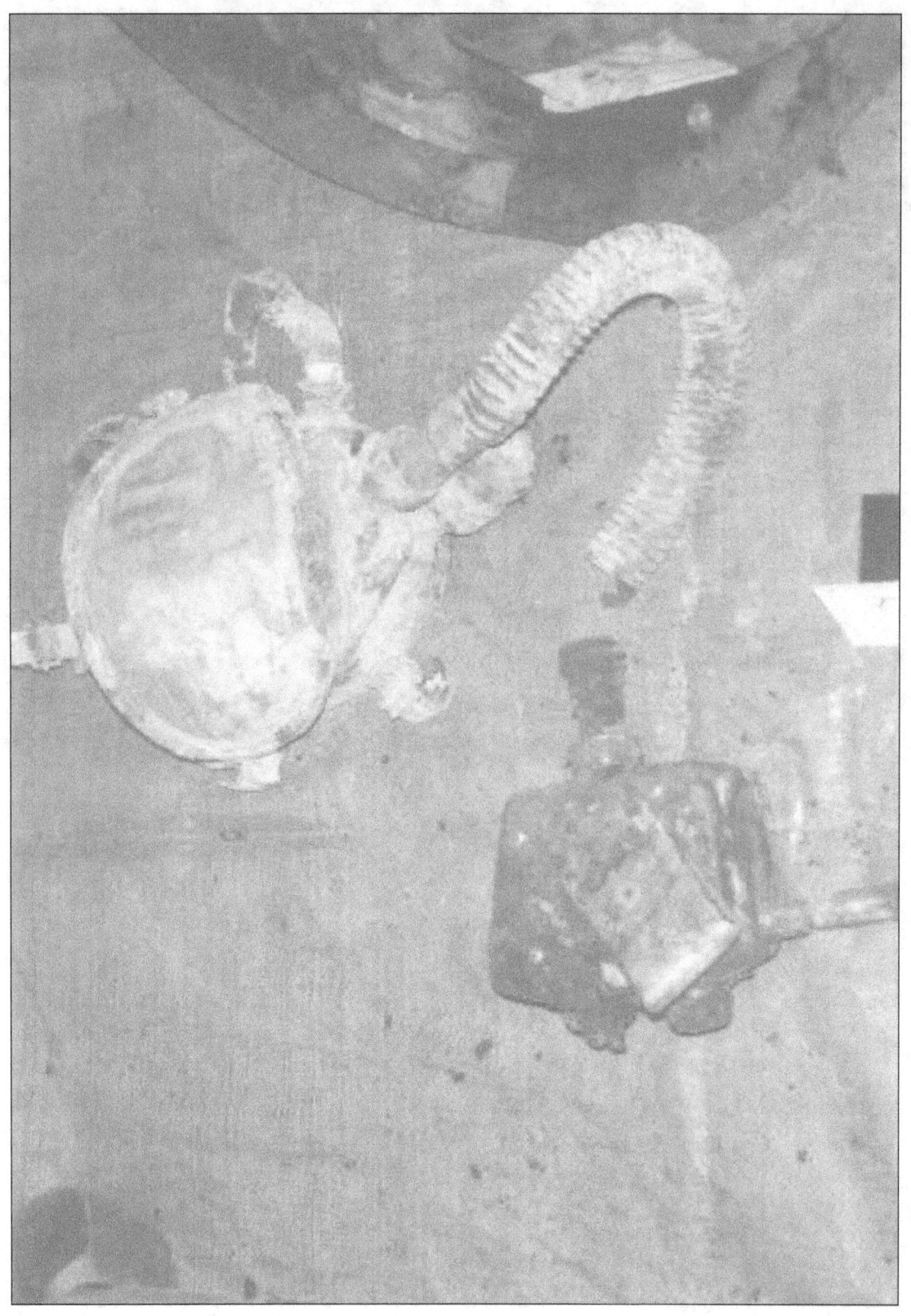

Photo by J. Gordon Routley

Facepiece from Scott 2A breathing apparatus showing failure of the low pressure breathing tube at the regulator and complete coverage of the lens with sodium residue.

Appendix C (continued)

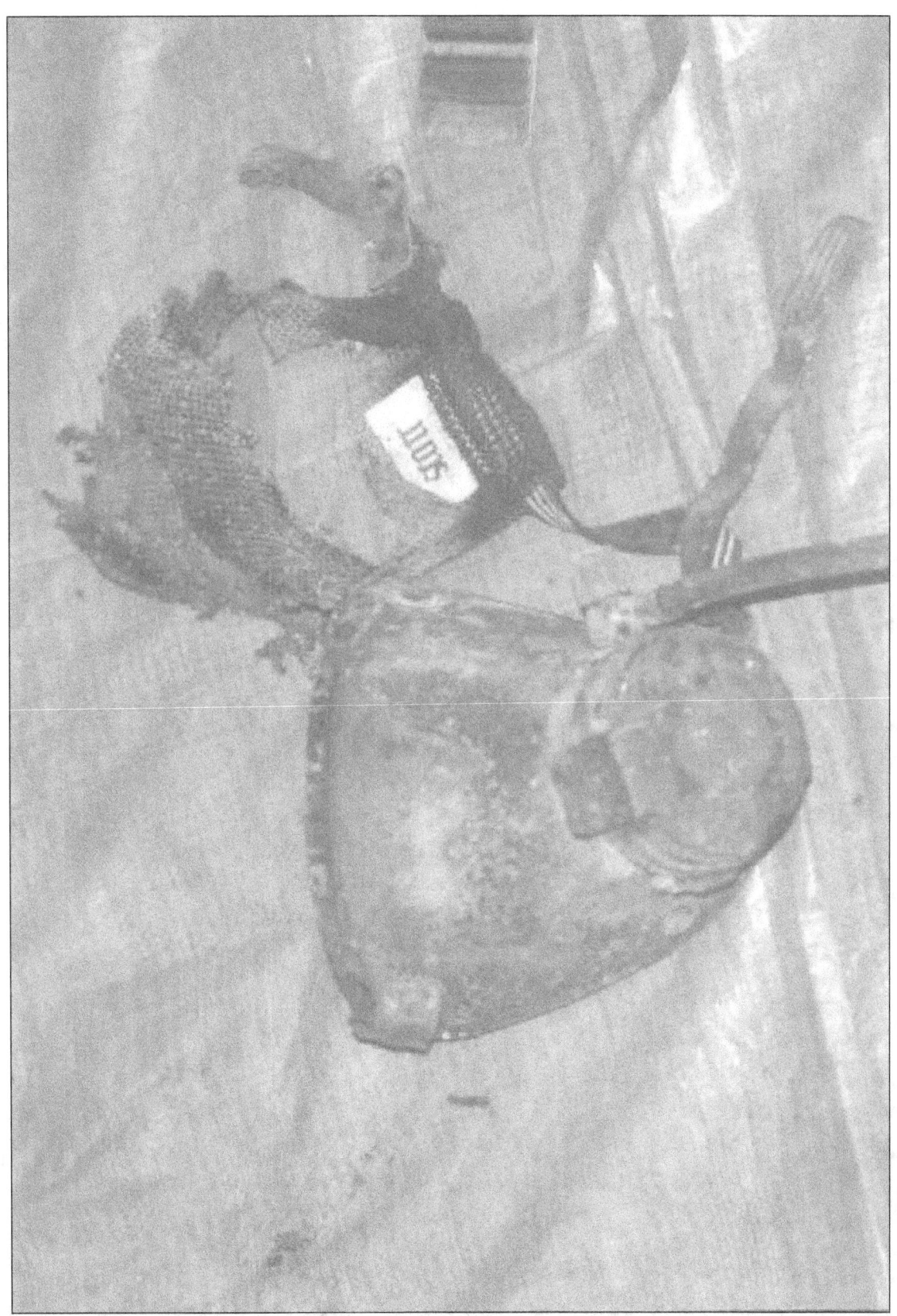

Photo by J. Gordon Routley

Outside view of Scott 4.5 facepiece showing complete coverage of lens with metallic sodium and residue.

Appendix C (continued)

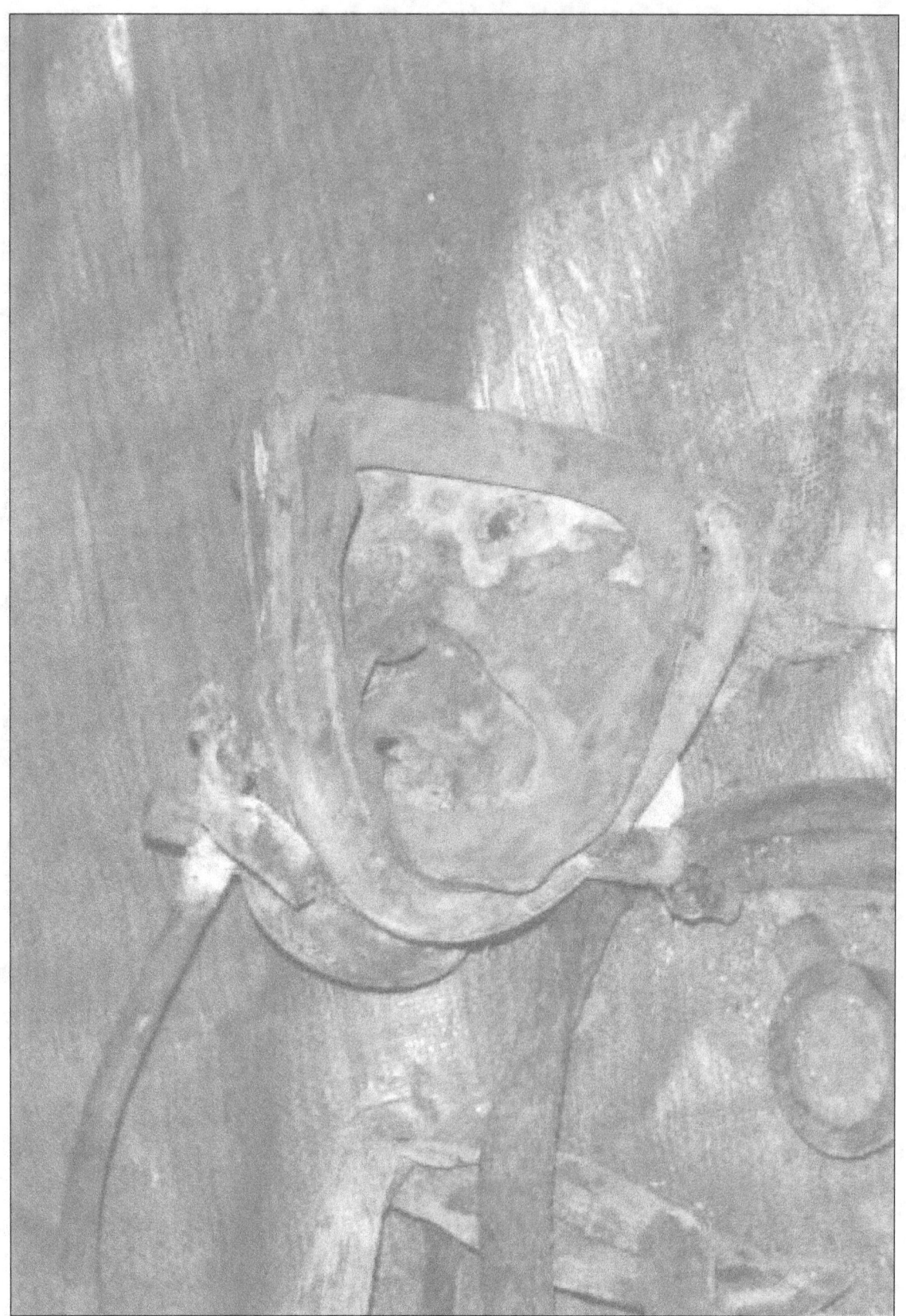

Photo by J. Gordon Routley

Inside view of Scott 4.5 facepiece showing complete coverage of lens and penetration of metallic sodium through the lens.

Appendix C (continued)

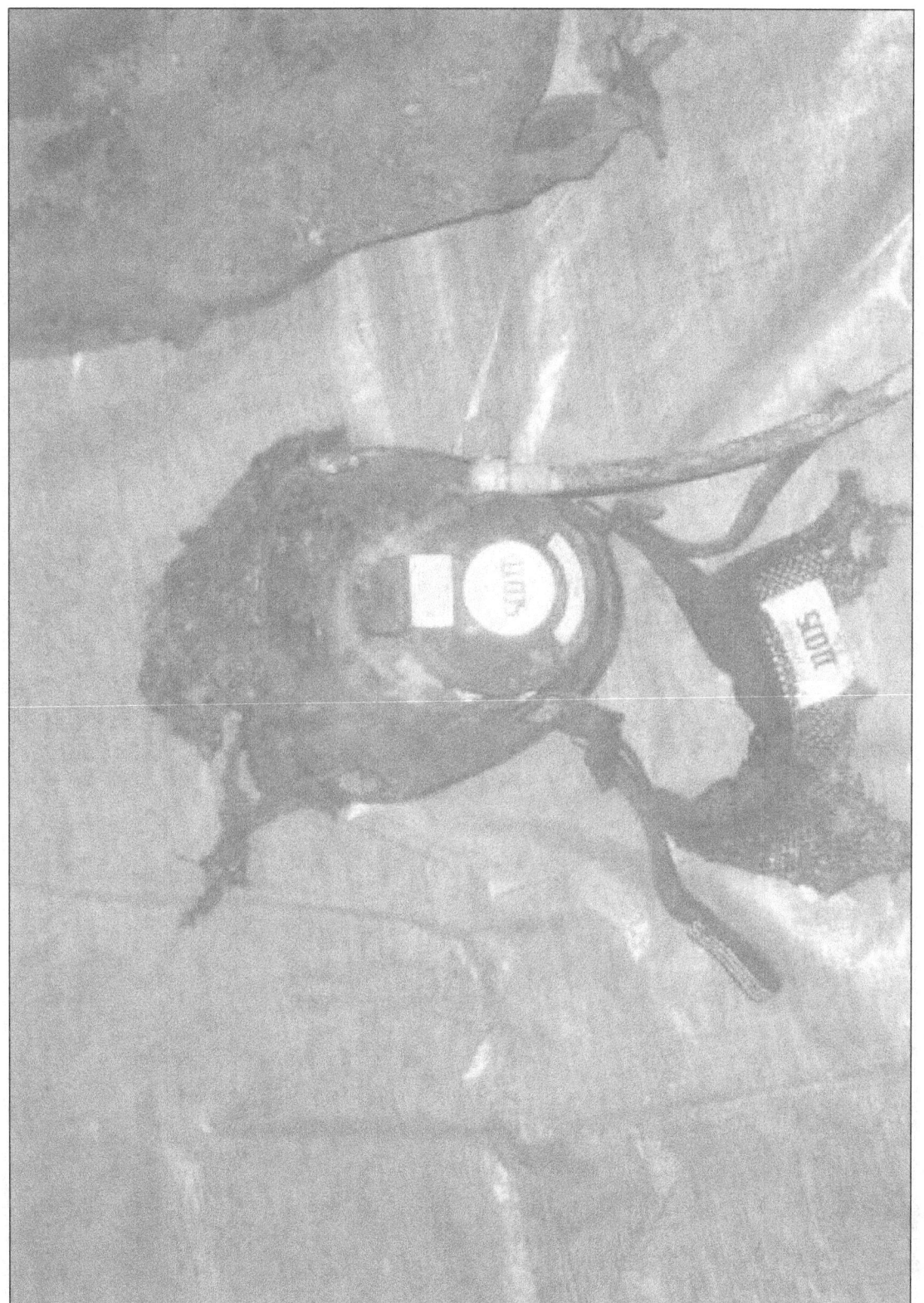

Scott facepiece showing damage caused by splatter of sodium on the lens and failure of the net retaining systems that secures the facepiece to the user's head.

Photo by J. Gordon Routley

Appendix C (continued)

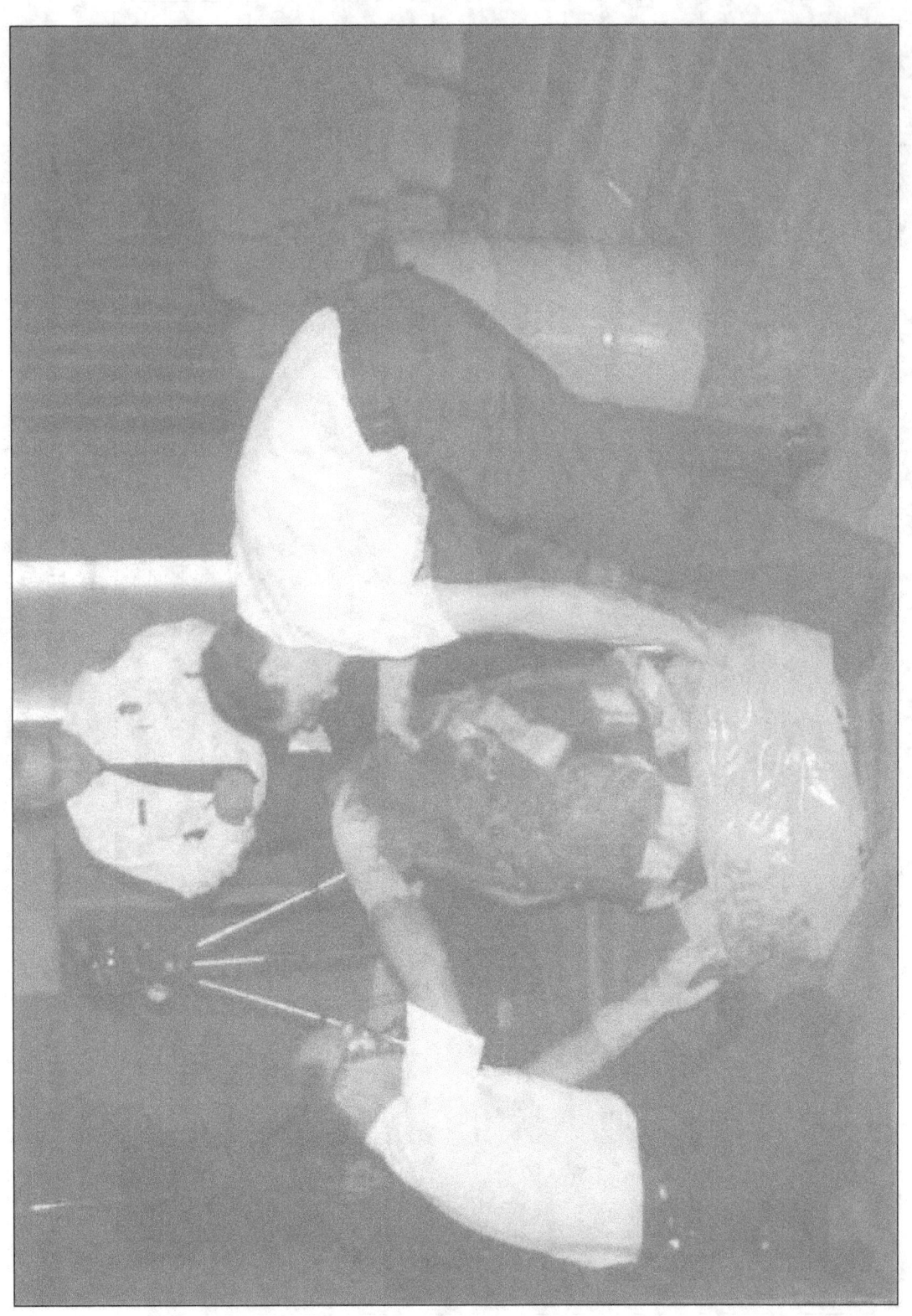

Photo by J. Gordon Routley

Removal of protective clothing from storage container. All of the protective clothing that was worn by the injured members, as well as their station uniforms and breathing apparatus, were impounded and stored in recovery drums for later examinations.

Appendix C (continued)

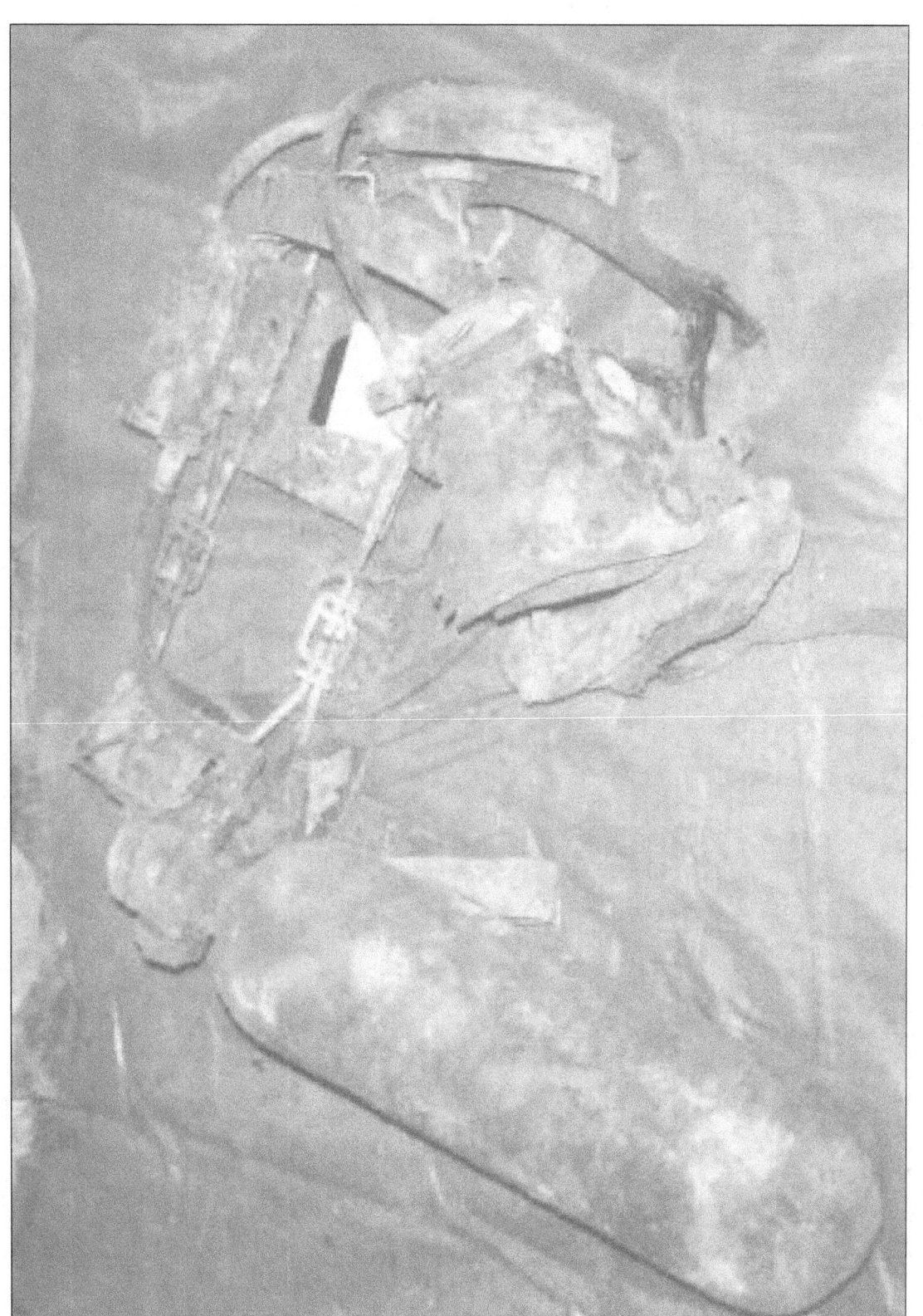

Photo by J. Gordon Routley

Scott 4.5 breathing apparatus showing major damage to facepiece lens and facepiece mounted regulator caused by direct contact with molten sodium. The metal penetrated the plastic lens and burned through the net and take-up strap that secured the facepiece to the users' head.

Appendix C (continued)

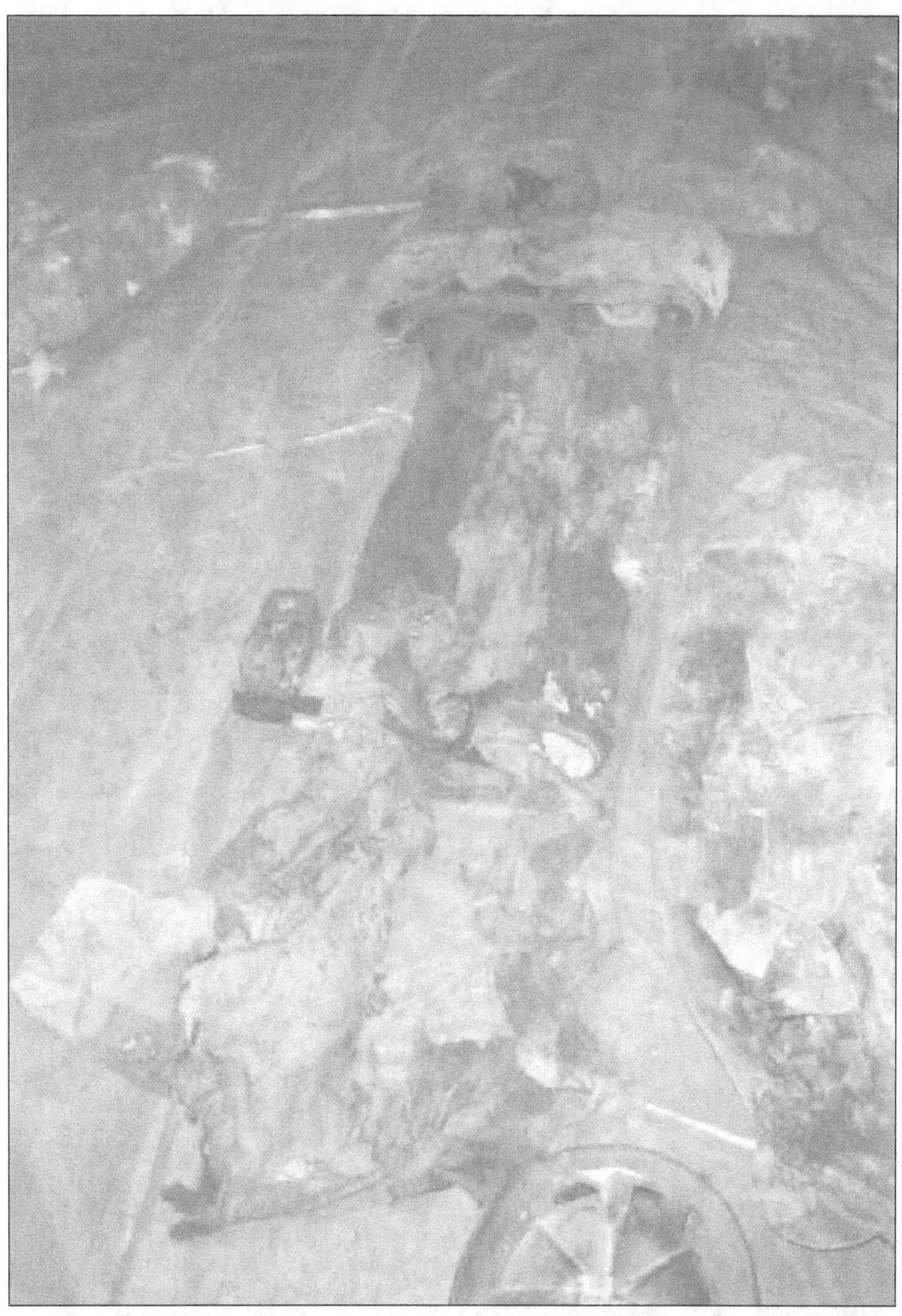

Photo by J. Gordon Routley

Remnants of non-fire retardant station uniform worn by critically burned lieutenant. The 3/4 length boots were worn turned down as show in the photo.

Appendix C (continued)

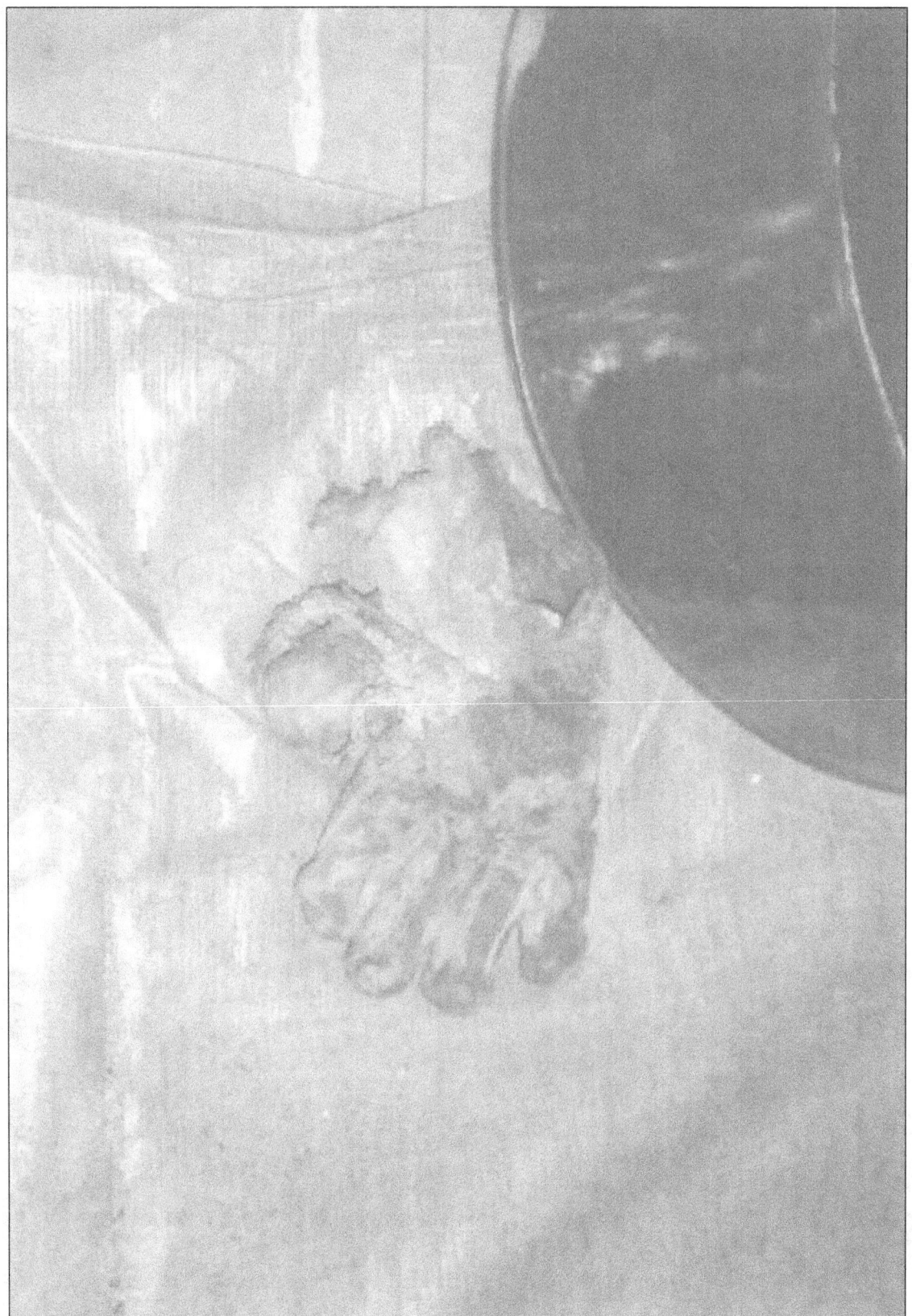

Photo by J. Gordon Routley

Leather firefighting glove that was contacted by molten sodium indicates severe damage.

Appendix C (continued)

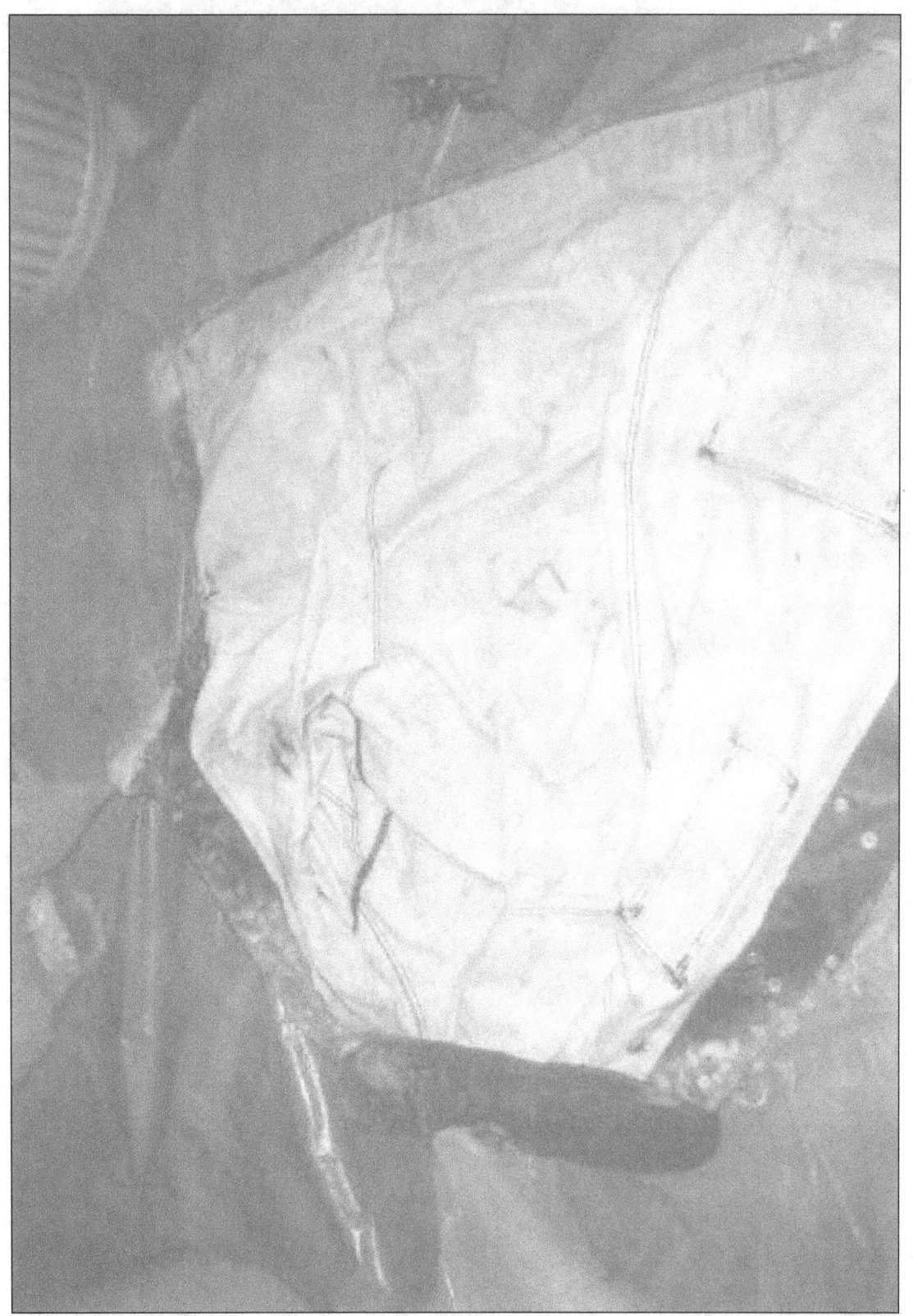

Photo by J. Gordon Routley

Coat worn without thermal liner was burned through in only a small area of the right arm pit. The firefighter wearing this coat would have been very critically burned if the amount of sodium splashed on the outer shell had been greater.

Appendix C (continued)

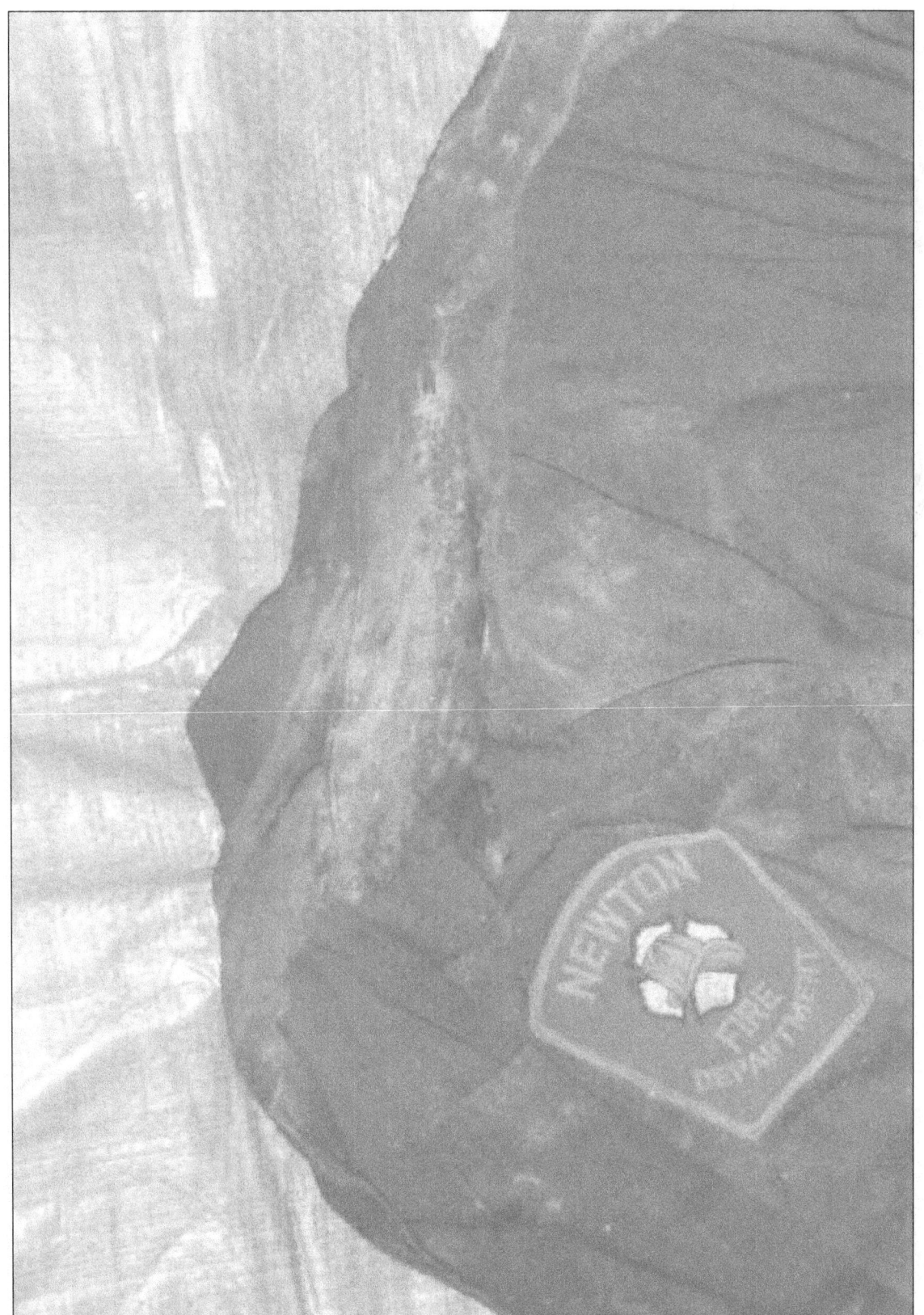

Photo by J. Gordon Routley

Non-FR station uniform shirt is burned in shoulder and collar area.

Appendix C (continued)

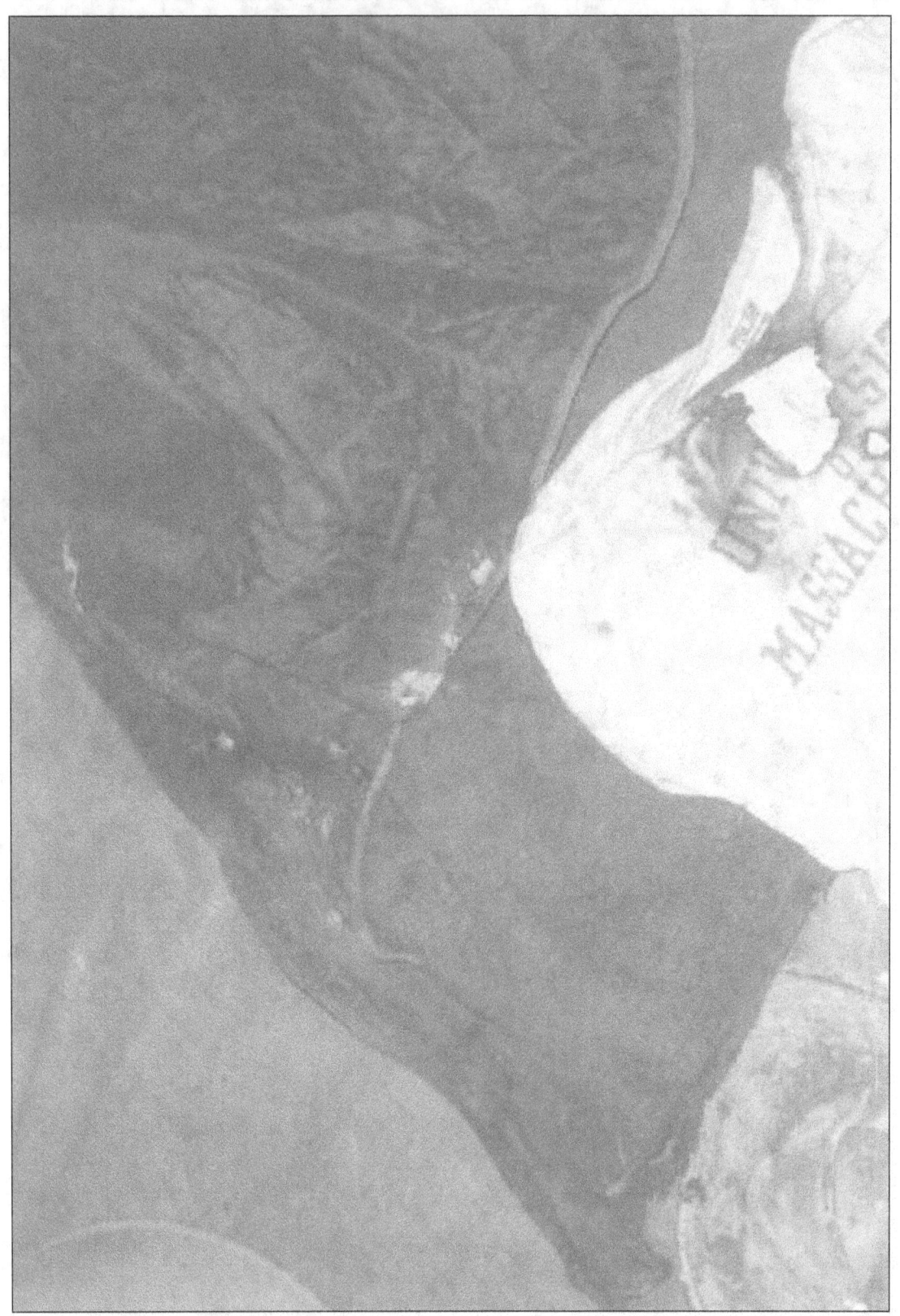

Winter liner (non-FR material) was ignited at lower edge, but burned in only a small area. The burns on the tee shirt indicated that penetration occurred in the collar area.

Photo by J. Gordon Routley

Appendix C (continued)

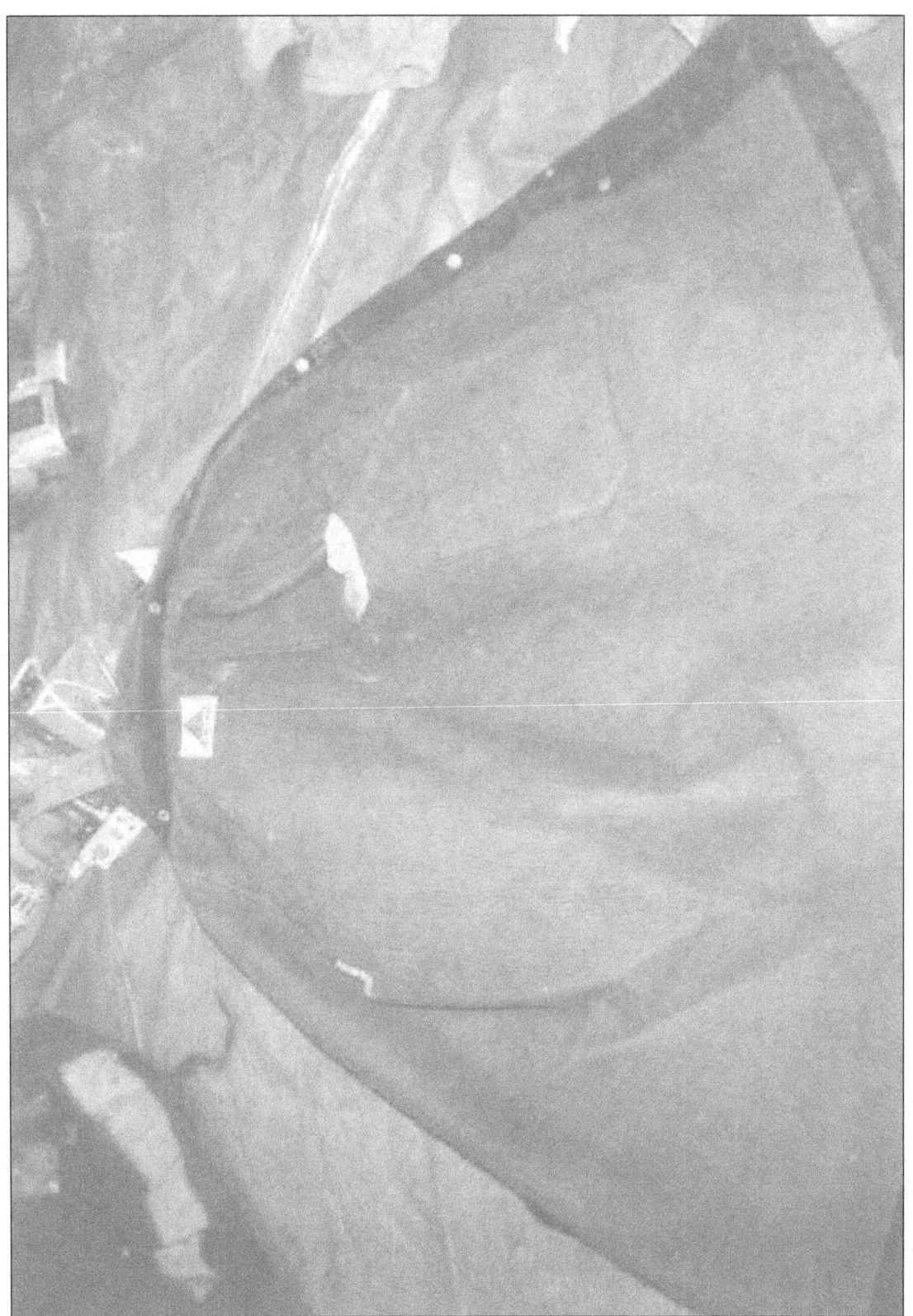

Photo by J. Gordon Routley

Inner view of coat shown in previous photograph shows that the molten sodium did not penetrate through the thermal liner except for a few small points.

Appendix C (continued)

Photo by J. Gordon Routley

Outer shell of turnout coat shows several areas where sodium burned through the outer shell; however, it did not penetrate Nomex®/neoprene thermal liner except in very small areas (shown in next photograph). The heat was sufficient to cause serious burns to the member who was wearing the coat.

Appendix C (continued)

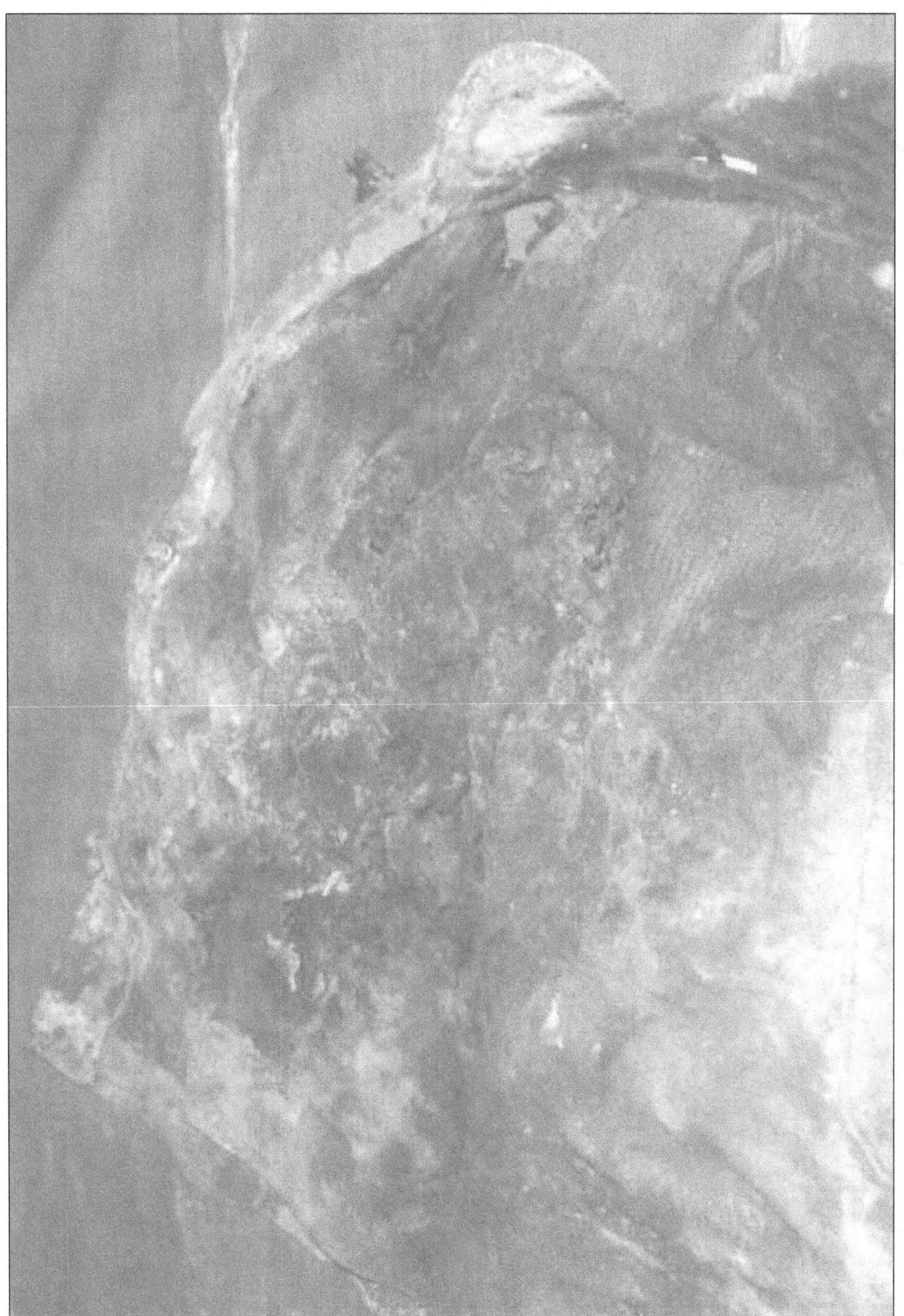

Photo by J. Gordon Routley

Inside view of coat shown in previous photograph shows major damage to thermal liner caused by molten sodium. The lieutenant who was wearing this coat received critical burn injuries.

Appendix C (continued)

Photo by J. Gordon Routley

Turnout coat worn by lieutenant who was critically burned. Outer shell and thermal liner were almost completely destroyed by direct contact with molten sodium.

Appendix C (continued)

Photo by J. Gordon Routley

Outer shell of turnout coat shows burn through to thermal liner in several areas on left side of upper body.

Appendix C (continued)

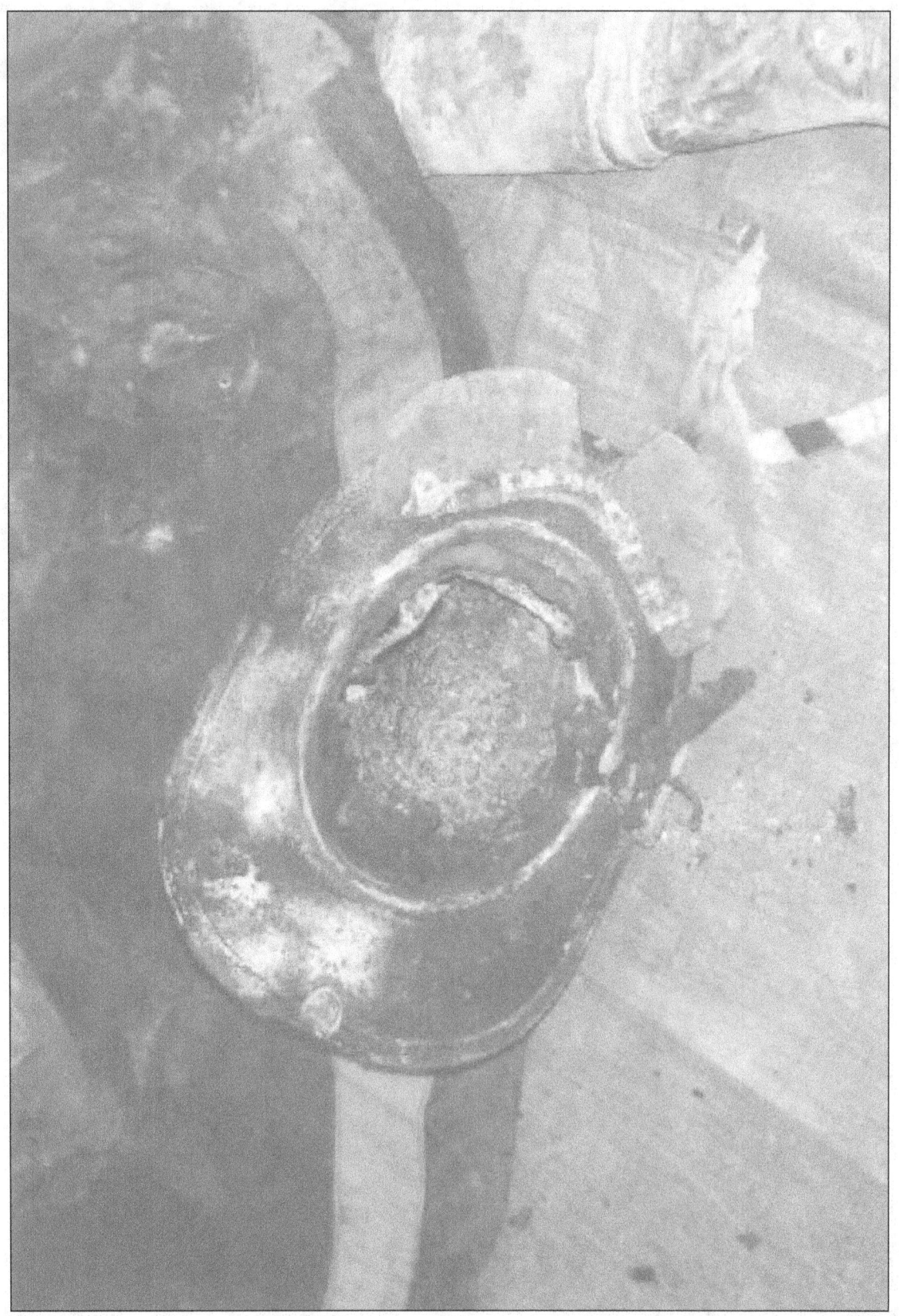

Photo by J. Gordon Routley

Inside suspension and earflaps of leather helmet were ignited and destroyed.

Appendix C (continued)

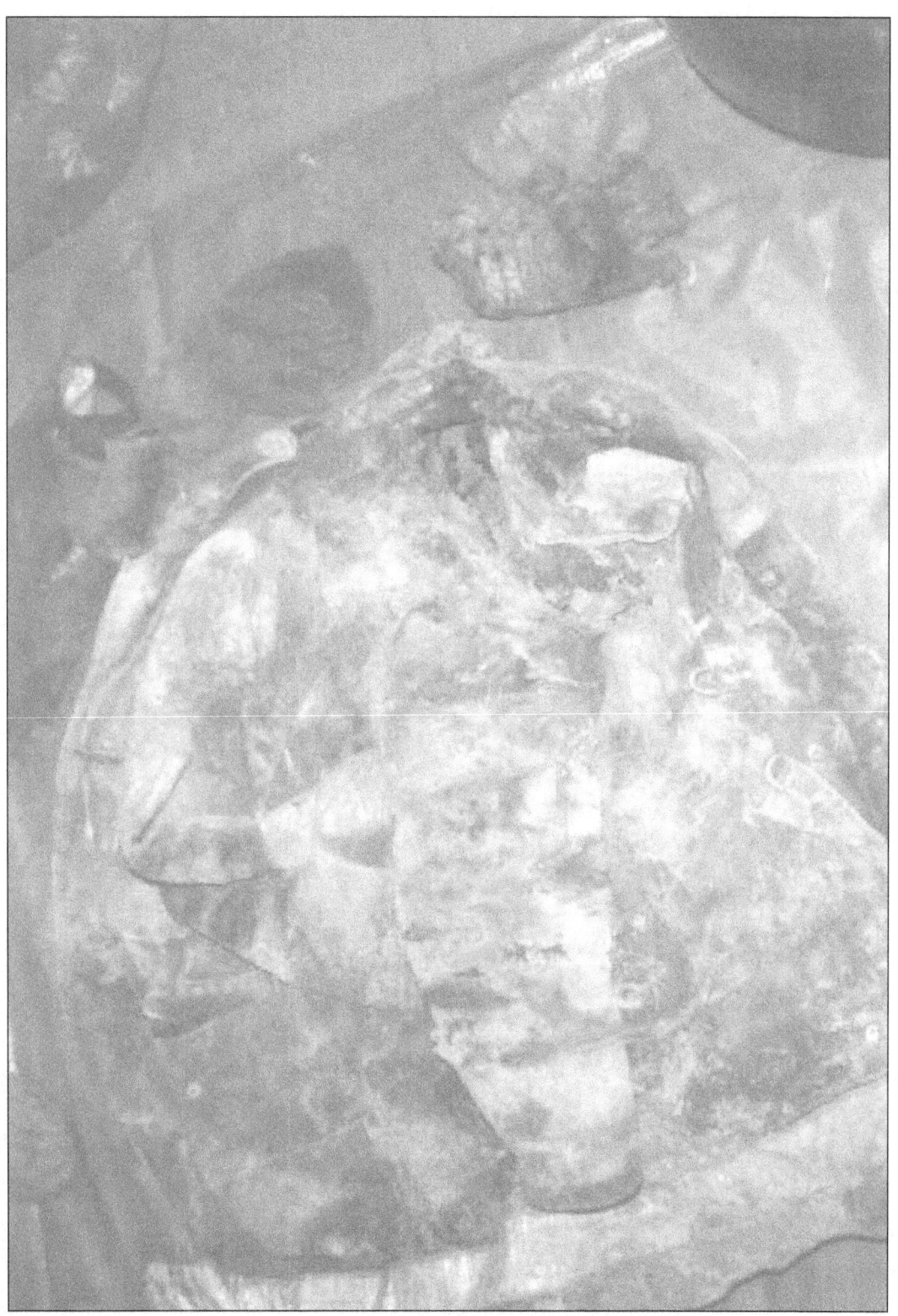

Photo by J. Gordon Routley

Turnout coat shows major damage to the right side caused by a splash of molten sodium. The molten metal penetrated the outer shell in the area of the right shoulder and ran down the inside of the coat, destroying the thermal liner. The right sleeve is also burned through in several locations.